D&D

STANDARD OIL ABBREVIATOR

Third Edition

Compiled by
the Association of
Desk and Derrick Clubs

D1225864

PennWell Books
PennWell Publishing Company
Tulsa, Oklahoma USA

Copyright © 1986 by
PennWell Publishing Company
1421 South Sheridan Road/P.O. Box 1260
Tulsa, Oklahoma 74101.

Library of Congress cataloging in publication data

D & D standard oil abbreviator.

1. Petroleum industry and trade—Abbreviations.
2. Gas industry—Abbreviations. I. Association of
Desk and Derrick Clubs. II. Title: D and D standard
oil abbreviator.

TN865.D2 1986   338.2′72820148   85-25895
ISBN 0-87814-299-1

Printed in the United States of America

1 2 3 4 5   90 89 88 87 86

Preface to the Third Edition

The *D&D Standard Oil Abbreviator* has become an indispensible tool within the oil and gas industry. The information it contains has made writing tasks much easier and has added uniformity.

When this revision was announced, the members of Desk and Derrick, as usual, came through with updated information from every aspect of the oil and gas industry. I have come to the conclusion that "we can abbreviate *everything*" if we put our minds to it. This Third Edition contains over 10,500 abbreviations and definitions for our industry.

Two new sections have been added. Within the Logging Section a new grouping was added for miscellaneous terms. Some of the entries are repeats from within the company groups, but this section also contains additional terms, thus alphabetizing all logging terms together. A new section was added—"Pipe Coating Terminology and Definitions"—and I am sure this will be a growing addition.

These abbreviations and terms have been mechanically alphabetized through the IBM 5520 System, which takes into account symbols, i.e., hyphens, slashes, etc., and arranges the abbreviations and terms accordingly. Once you use the Abbreviator, you'll appreciate the ease of using this system.

I wish to thank all the members of D&D who made contributions to this new edition and also those people who helped arrange this information for publication.

My wish for this edition is to further carry on the hope of the first *D&D Standard Oil Abbreviator*–to fill a need.

Linda d'Allessandro Weatherly

WHAT IS D&D?

Conceived by an ambitious secretary, nurtured by women with vision, encouraged by cooperative employers and a progressive industry—this is the history of Desk and Derrick. It is a unique organization of nine thousand women employed in the petroleum and allied industries. All 123 clubs, spread over the North American continent, are dedicated to the proposition that greater knowledge of the petroleum industry will result in greater service in job performance.

From New Orleans, where the first club was formed in April 1949, the idea spread to Jackson, Mississippi, Los Angeles, and Houston; in September 1962 in Houston the "Queen of Clubs" held its first annual convention.

The Desk and Derrick purpose is implemented by educational monthly programs, by field trips to industry installations, and by special study courses.

Nonshareholding, noncommercial, nonprofit, nonpartisan, and nonbargaining in its policies, the organization nevertheless has very positive concepts on the value of education for women.

CONTENTS

A

A	abstract (i.e., A-10)
A	angstrom unit
A-Cem	acoustic cement
A&A	adjustments and allowances
A/	acidized with
A/C	air conditioning
A/CLR	air cooler
A/R	accounts receivable
AA	after acidizing, as above
AAR	Association of American Railroads
ABC	Audit Bureau of Circulation
abd	abandoned
abd loc	abandoned location
abd-gw	abandoned gas well
abd-ow	abandoned oil well
abdogw	abandoned oil & gas well
ABHL	absolute bottom-hole location
ABM	Atlas Bradford modified
abrsi jet	abrasive jet
abs	absolute
ABS	acrylonitrile butadiene styrene rubber
absrn	absorption
abst	abstract
abt	about
abun	abundant
abv	above
ac	acid
ac	acre(s), acreage
AC	alternating current

1

AC	Austin chalk
ac-ft	acre-feet
ACC	access
ACCEL	accelerometers
ACCESS	accessory, accessories
acct	account (ing)
accum	accumulative, accumulator
acd	acidize (ed) (ing)
acfr	acid fracture treatment
ac-ft	acre feet
ACLD	air cooled
ACLR	air cooler
ACM	acid-cut mud
acrg	acreage
ACS	American Chemical Society
ACSR	aluminum conductor steel reinforced
ACT	actual
ACT	actuated, actuator
ACT	automatic custody transfer
ACW	acid-cut water
AD	actual drilling
AD	authorized depth
ADC	actual drilling cost
add	additive
addl	additional
ADH	adhesive
adj	adjustable
adm	administration, administrative
ADOM	adomite
ADP	automatic data processing
adpt	adapter
adspn	adsorption
ADT	actual drilling time
advan	advanced
AER	aeration, aerator
AF	acid frac
AF	after fracture
AF/CLR	after cooler

AF/COND	after condenser
AFC	Authorization for Commitment
AFC	Authorized for Construction
AFD	auxiliary flow diagram
AFE	Authorization for Expenditure
affd	affirmed
afft	affidavit
AFIT	after federal income tax
AFP	average flowing pressure
AFRA	average freight rate assessment
AG	agitator
AGGR	aggregate
aglm	agglomerate
AIR	average injection rate
AIR COND	air conditioning
AJT	actual jetting time
AL	aluminum
AL	artificial lift
Alb	Albany
alc	alcoholic
ALCOA	Aluminum Company of America
alg	algae
alg	along
ALIGN	alignment (ing)
alk	alkaline, alkalinity
alkyl	alkylate, alkylation
ALLOW	allowable, allowance
alm	alarm
ALOC	allocation
alt	alternate
ALT	altitude
ALY	alloy
amb	ambient
AMI	area of mutual interest
AMM	ammeter
amor	amorphous

amort	amortization
AMP	American melting point
amp	ampere
amp hr	ampere hour
amph	amphipore
Amph	Amphistegina
AMR	addition or modification request
AMR	amount not reported
amt	amount
an	annulus
anal	analysis, analytical
ANC	Anchor (age)
ang	angle, angular
Angul	Angulogerina
anhy	anhydrite, anhydritic
anhyd	anhydrous
Ann	annulus
ANR	amount not reported
ANS	Alaskan North Slope
ANUB	annubar
ANUC	annunciator
ANYA	allowable not yet available
AOAC	Association of Official Agricultural Chemistry
AOF	absolute open flow potential (gas well)
AP&VMA	American Paint and Varnish Manufacturers Association
APHA	American Public Health Association
API	American Petroleum Institute
app	appears, appearance
APPAR	apparatus
appd	approved
appl	appliance
appl	applied
applic	application

approx	approximate (ly)
apr	apparent (ly)
APR	average penetration rate
apt	apartment
aq	aqueous
AQCR	air quality control region
AQMA	air quality maintenance area
AR	acid residue
Ara	Arapahoe
arag	aragonite
Arb	Arbuckle
arch	architectural
Archeo	Archeozoic
aren	arenaceous
arg	argillaceous
arg	argillite
ark	arkose(ic)
Arka	Arkadelphia
arm	armature
arnd	around
ARO	after receipt of order (purchasing term)
ARO	at rate of
arom	aromatics
ARR	arrange (ed) (ing) (ment)
AS	after shot
AS	anhydrite stringer
AS&W ga	American Steel & Wire gauge
ASA	American Standards Association
ASAP	as soon as possible
asb	asbestos
asbr	absorber
ASD	abandoned–salvage deferred
asgmt	assignment
Ash	ashern
ASO	acid-soluble oil

asph	asphalt, asphaltic
assgd	assigned
assn	association
assoc	associate (d) (s)
asst	assistant
assy	assembly
ASTM	American Society for Testing & Materials
astn	asphaltic stain
ASW	adjustable spring wedge
AT	acid treat (ment)
AT	after treatment
AT	all thread
At	Atoka
at	atomic
at wt	atomic weight
ATC	after top center
ATD	approved total depth
ATF	automatic transmission fluid
atm	atmosphere, atmospheric
ATP	Authorization to Proceed
ATP	average treating pressure
ATP	average tubing pressure
ATT	attach (ed) (ing) (ment)
att	attempt(ed)
atty	attorney
aud	auditorium
Aus	Austin
auth	authorized
auto	automatic
auto	automotive
autogas	automotive gasoline
aux	auxiliary
AV	annular velocity
AV	Aux Vases sand
av	aviation
avail	available
AVC	automatic volume control
avg	average

avgas	aviation gasoline
AW	acid water
AWD	award
AWG	American Wire Gauge
awtg	awaiting
AWWA	American Water Works Association
az	azimuth
aztrop	Azeotropic

————— **B** —————

B	billion
B	bulletin
B & B	bell and bell
B & CB	beaded and center beaded
B & S	bell and spigot
B Hn	Big Horn
B slt	base of the salt
B. In.	Big Injun
B. Ls	Big Lime
B. Riv	Black River
B. Bl	Base Blane
B.E.	bevelled end
BS&W	basic sediment and water
B&B	bent & bowed pipe
B&F	bell and flange
B&S	Brown & Sharpe (gauge)
B/	base
B/	bottom of given formation (i.e., B/Frio)
B/B	back to back
B/B	barrels per barrel
B/D	barrels per day
B/dry	bailed dry
B/hr	barrels per hour
B/JT	ball joint

B/L	bill of lading
B/M	bill of material
b/off	buck-off
b/on	buck-on
B/S	back scuttled
B/S	base salt
B/S	bending schedule
B/S	bill of sale
B/SD	barrels per stream day (refinery)
B/VESS	bulk vessel
B/Vlv	ball valve
BA	barrels of acid
bail	bail (ed)
BAL	balance
Ball	Balltown sand
bar	barite (ic)
Bar	Barlow Lime
bar	barometer, barometric
BAR	barrels acid residue
Bark Crk	Barker Creek
Bart	Bartlesville
base	basement (granite)
bat	battery
BAT	before acid treatment
Bate	Bateman
BAW	barrels acid water
BAWPD	barrels acid water per day
BAWPH	barrels acid water per hour
BAWUL	barrels acid water under load
BB	bridged back
BB fraction	butane-butene fraction
BBE	bevel both ends
bbl	barrel
BC	barrels of condensate
Bcf	billion cubic feet
Bcfd	billion cubic feet per day
BCPD	barrels condensate per day
BCPH	barrels condensate per hour

BCPMM	barrels condensate per million
BD	barrels of distillate
bd	board
BD	budgeted depth
bd ft	board foot (feet)
BD-MLW	barge deck to mean low water
Bd'A	Bois d'Arc
BDA	breakdown acid
BDF	broke (break) down formation
BDL	bundle
BDNG	bedding
BDO	barrels diesel oil
BDP	breakdown pressure
BDPD	barrels distillate per day
BDPH	barrels distillate per hour
BDT	blow-down test
Be	Baumé
Be	Berea
be	box end
Bear Riv	Bear River
bec	becoming
Beck	Beckwith
Bel	Beldon
Bel C	Belle City
Bel F	Belle Fourche
Belm	belemnites
Ben	Benoist (Bethel) sand
Ben	Benton
Bent	bentonite
berm	berm, sloped wall to keep out flooding
bev	bevel (ed)
BF	barrels fluid
BF	blind flange
bf	buff
BFIT	before federal income tax
BFL	baffle

BFO	barrels frac oil
BFPD	barrels fluid per day
BFPH	barrels fluid per hour
BFW	bailer feed water
BFW	barrels formation water
BFW	barrels fresh water
BFW	boiler feed water
BH	bottom hole
BHA	bottom-hole assembly
BHC	bottom-hole choke
BHCS	borehole compensated sonic
BHF	Bradenhead Flange
BHFP	bottom-hole flowing pressure
BHL	bottom-hole location
BHM	bottom-hole money
BHN	Brinell hardness number
BHO	bottom-hole orientation
BHP	bottom-hole pressure
bhp	brake horsepower
bhp-hr	brake horsepower-hour
BHPC	bottom-hole pressure closed (*see also* SIBHP and BHSIP)
BHPF	bottom-hole pressure flowing
BHPS	bottom-hole pressure survey
BHSIP	bottom-hole shutin pressure
BHT	bottom-hole temperature
BID SUM	bid summary
Big.	Bigenerina
Big. f.	Bigenerina floridana
Big. h.	Bigenerina humblei
Big. nod.	Bigenerina nodosaria
BIN	binary
bio	biotite
bit	bitumen
bit	bituminous
bkdn	breakdown

BKFLSH	back flush
bkr	breaker
BKWSH	backwash
BL	barrels load
bl	blue
BL&AW	barrels load & acid water
Bl/Cb	blast cabinet
BL/JT	blast joint
BLC	barrels load condensate
BLCPD	barrels load condensate per day
BLCPH	barrels load condensate per hour
bld	bailed
BLD FLG, BF	blind flange
bldg	bleeding
bldg	bleeding gas
bldg	building
bldg drk	building derrick
bldg rds	building roads
bldo	bleeding oil
bldrs	boulders
BLDWN	blowdown
BLE	bevel large end
blg	bailing
Blin	Blinebry
blk	black
blk	block
Blk Lf	Black Leaf
Blk Li	Black Lime
blk lnr	blank liner
BLND	blend (ed) (er) (ing)
BLO	barrels load oil
blo	blow
BLOPD	barrels load oil per day
BLOPH	barrels load oil per hour
BLOR	barrels load oil recovered
Blos	blossom
BLOTBR	barrels load oil to be recovered

BLOYTR	barrels load oil yet to recover
BLPD	barrels of liquid per day
blr	bailer
BLR	boiler
blts	bullets
BLW	barrels load water
BLWPD	barrels load water per day
BLWPH	barrels load water per hour
BLWR	blower
BLWTR	barrels load water to recover
BM	barrels mud
BM	benchmark
BM	Black Magic (mud)
BMEP	brake mean effective pressure
BMI	black malleable iron
bmpr	bumper
bn	brown
bnd	band (ed)
bndry	boundary
bnish	brownish
BNO	barrels new oil
BNW	barrels of new water
bnz	benzene
BO	backed out (off)
BO	barrels oil
BO	blew out
BO	blocked off
BO	free-point back off
BOCD	barrels oil per calendar day
BOCS	Basal Oil Creek sand
BOD	barrels oil per day
BOD	biochemical oxygen demand
Bod	Bodcaw
BOE	bevel one end
BOE	blowout equipment
Bol.	Bolivarensis

Bol. a.	Bolivina a.
Bol. flor.	Bolivina floridana
Bol. p.	Bolivina perca
Bonne	Bonneterre
BOP	blowout preventer
BOPD	barrels oil per day
BOPE	blowout preventer equipment
BOPH	barrels oil per hour
BOPPD	barrels oil per producing day
BOS	brown oil stain
bot	bottom
BP	back pressure
BP	base Pennsylvanian
BP	Bearpaw
BP	boiling point
BP	bridge plug
BP	bulk plant
BP	bull plug
BP Mix	butane and propane mix
BP/CLR	bypass cooler
BPD	barrels per day
BPH	barrels per hour
BPLO	barrels of pipeline oil
BPLOPD	barrels of pipeline oil per day
BPM	barrels per minute
BPSD	barrels per stream day
BPV	back-pressure valve
BPWPD	barrels per well per day
BR	building rig
BR	building road
brach	brachiopod
BRC	brace (ing) (ed)
brec	breccia
BRFL/V	butterfly valve
brg	bearing
Brid	bridger
brit	brittle

brk	break (broke)
brkn	broken
brkn sd	broken sand
BRKR	breaker
BRKS	brakes
brksh	brackish (water)
brkt(s)	brackets(s)
Brn Li	brown lime
brn or br	brown
brn sh	brown shale
Brom	bromide
brtl	brittle
bry	bryozoa
BS	ball sealers
BS	basic sediment
BS	Bone Spring
BS	bottom sediment
BS	bottom settlings
BS&W	bottom (basic) sediment & water
Bscf	billion standard cubic feet
Bscf/d	billion standard cubic feet per day
BSE	bevel small end
BSFC	brake specific fuel consumption
BSHG	bushing
BSI	British Standards Institution
bskt	basket
bsl	basal
bsmt	basement
BSPL	base plate
BSTR	booster
BSUW	black sulfur water
BSW	barrels salt water
BSWPD	barrels salt water per day
BSWPH	barrels salt water per hour
BT	Benoist (Bethel) sand
BTC	buttress thread coupling
BTDC	before top dead center

BTFL/V	butterfly valve
btm (d)	bottom (ed)
btm chk	bottom choke
btry	battery
BTU	British thermal unit
btw	between
BTWLD	butt weld
BTX	benzene toluenexylene (unit)
bu	bushel
Buck	Buckner
Buckr	buckrange
Bul. text.	Buliminella textularia
Bull W	Bullwaggon
Bum.	bottom-hole pressure bomb
bunr	burner
Burg	Burgess
butt	buttress thread
BUZ	buzzer
BV	block valve
BV/WLD	beveled for welding
BW	barrels of water
BW	boiled water
BW	butt weld
BW ga	Birmingham (or Stubbs) iron wire gauge
BW/D	barrels of water per day
Bwg	Birmingham wire gauge
BWL	barrel water load
BWL	body wall loss
BWOL	barrels water over load
BWPD	barrels water per day
BWPH	barrels water per hour
bx	box (es)
BYP	bypass

——————— **C** ———————

C	Celsius
C	center (land description)

C	centigrade
c	coarse (ly)
C	core hole
C & F	cost and freight
C & P	cellar & pits
C & W	coat and wrap (pipe)
C to C	center to center
C to E	center to end
C to F	center to face
C.I.F.	cost insurance and freight
C.O.P.	completed on pump
C&A	compression and absorption plant
C&C	circulate & condition
C&CH	circulated and conditioned hole
C&CM	circulated and conditioned mud
C&R	circulate and reciprocate
C/	contractor (i.e., C/John Doe)
C/A	commission agent
C/BM	crawl beam
C/H	cased hole
C/L	center line
c/o	care of
C/W	complete with
CA	corrosion allowance
CAB	cabinet
CaCl2	calcium chloride
Cadd	Caddell
CAG	cut across grain
cal	calcite, calcitic
cal	caliche
CAL	caliper log
cal	caliper survey
cal	calorie
Calc	calcareous, calcerenite
Calc	calcium
Calc	calculate (ed), calculation
calc gr	calcium-base grease

calc	calceneous
CALIBR	calibrate, calibration
Calv	Calvin
Camb	Cambrian
Cycl canc.	Cyclamina cancellata
Cane Riv	Cane River
Cany	canyon
CaO	calcium oxide
CAOF	calculated absolute open flow
cap	capacitor
cap	capacity
Cap	Capitan
Car	Carlile
carb	carbonaceous
carb tet	carbontetrachloride
Carm	Carmel
Casp	Casper
CAT	carburetor air temperature
Cat	Catahoula
CAT	catalog
CAT	catalyst, catalytic
cat ckr	catalytic cracker
Cat Crk	Cat Creek
cath	cathodic
caus	caustic
cav	cavity
CB	changed (ing) bits
CB	continuous blowdown
CB	core barrel
CB	counterbalance (pumping equip.)
CBC	cement dump bailer
CBL	cable (ing)
CBU	circulate bottoms up
CC	calcium chloride
CC	carbon copy
CC	casing cemented (depth)
CC	closed cup
cc	cubic centimeter

C-Cal	contact caliper
ccBU	circulate bottoms up
CCHF	center of casinghead flange
Cck	casing choke
CCL	casing collar locator
CCLGO	cat-cracked light gas oil
CCM	condensate-cut mud
CCP	central compressor plant
CCP	critical compression pressure
CCPR	casing collar perforating record
CCR	Conradson carbon residue
CCR	critical compression ratio
CCS	California Coordinate System
CS	cast carbon steel
CCS	computer control system
CCU	catalytic cracking unit
ccw	counterclockwise
CD	calendar day
CD	cold drawn
CD	contract depth
CD PL	cadmium plate
CDB	cement dump bailer
CDB	common data base
CDBTF	common data base task force
CDL	cut drilling line
CDM	continuous dipmeter survey
CDO	certified drawing outline
Cdr Mtn	Cedar Mountain
CDS	continuous directional service
cdsr	condenser
Cdy	Cody (Wyoming)
cell	cellar
cell	cellular
cem	cement (ed)
CEMF	counter electromotive force

Ceno	Cenozoic
cent	centralizers
centr	centrifugal
ceph	cephalopod
Cert. ex.	Ceratobulimina eximia
CERT	certified
CET	cement evaluation
CF	casing flange
CF	clay filled
Cf	Cockfield
CF	cold finished
cf	cubic foot (feet)
CFB & G	companion flange bolt and gasket
CFBO	companion flanges bolted on
cfd	cubic feet per day
CFE	contractor furnished equipment
cfg	cubic feet gas
cfgd	cubic feet gas per day
cfgh	cubic feet gas per hour
CFM	continuous flowmeter
cfm	cubic feet per minute
CFOE	companion flange one end
cfp	cubic feet per pound
CFR	cement friction reducer
CFR	cement friction retarder
CFRC	Coordinating Fuel Research Committee
cfs	cubic feet per second
CG	center of gravity
cg	centigram
cg	coarse grained
cg	coring
CG	corrected gravity
cglt	conglomerate, conglomeritic
C-gr	coarse grained
cgs	centimeter-gram-second system

CH	casinghead (gas)
ch	chert
ch	choke
CH	closed hole
CH	core hole
CH OP	chain operated
chal	chalcedony
CHAM	chamfer
Chapp	Chappel
CHAR	characteristics
Char	Charles
Chatt	Chattanooga shale
CHD	closed hydrocarbon drain
chem	chemical, chemist, chemistry
chem prod	chemical products
Cher	Cherokee
Ches	Chester
CHF	casinghead flange
CHG	casinghead gas
chng	change (ed) (ing)
Chngd DP	changed drillpipe
chrg	charge (ed) (ing)
Chim H	Chimney Hill
Chim R	Chimney Rock
Chin	Chinle
chit	chitin (ous)
chk	chalk
chk	choke
Chkbd	checkerboard
chkd	checked
CHKD PL	checkered plate
CHKV	check valve
chky	chalky
chl	chloride (s)
chl	chloritic
chl log	chlorine log
CHLR	chlorinator
CHMBR	chamber
CHNL	channel

Chou	Chouteau lime
CHP	casinghead pressure
chrm	chairman
chromat	chromatograph
chrome	chromium
cht	chart
cht	chert
chty	cherty
Chug	chugwater
CI	cast iron
CI	contour interval (map)
CI engine	compression-ignition engine
Cib.	Cibicides
Cib. h.	Cibicides hazzardi
CIBP	cast-iron bridge plug
CIE	crude industrial ethanol
Cima	Cimarron
CIP	cement in place
CIP	closed-in pressure
cir	circle
cir	circuit
cir	circular
cir mils	circular mils
circ	circulate, circulating, circulation
Cis	Cisco
ck	cake
ck	check
CK Mtn	Cook Mountain
cksn	chicksan
CL	carload
cl	centiliter
CL	class
Clag	Clagget
Claib	Claiborne
Clarks	Clarksville
clas	clastic
CLASS	classification
Clav	Clavalinoides

Clay	Clayton
Clay	Claytonville
Cleve	Cleveland
Clfk	Clearfork
CLFR	clarifier
CLG	cooling
CLG/TWR	cooling tower
Cliff H	Cliff House
CLKG	caulking
CLMP	canvas-lined metal petal basket
cln (d) (g)	clean (ed) (ing)
Clov	cloverly
clr	clear, clearance
CLR	cooler
CLR/TWR	cooling tower
clrg	clearing
clsd	closed
CLTR	collector
Clyst	Claystone
Cl_2	chlorine
cm	centimeter
cm/sec	centimeters per second
CMA	acoustic caliper
CMC	sodium carboxymethylcellulose
Cmchn	comanchean
CMPARTR	comparator
CMPD	compound
Cmpt	compact
cmt(d)(g)(r)	cement (ed) (ing) (er)
CN	cetane number
CN/BD	control building
cncn	concentric
CND	conduit
CNL	compensated neutron log
CNR	corner
cntf	centrifuge
cntl	control (s)
CNTN	containment

cntr	center (ed)
cntr	container
cntr	controller
CNTWT	counter weight
Cnty	county
cnvr	conveyor
CO	carbon monoxide
CO	carbon oxygen
CO	circulated out
CO	clean out
CO	cleaning out, cleaned out
Co	company
CO	crude oil
CO & S	clean out & shoot
Co. Op.	company operated
Co. Op. S.S.	company-operated service stations
co-op	cooperative
COBOL	Common Business-Oriented Language
COC	Cleveland open cup
Coco	Coconino
COD	chemical oxygen demand
Cod	Codell
coef	coefficient
COF	calculated open flow (potential)
COG	coke oven gas
COH	coming out of hole
COL	collar
COL	colored
COL	column
Col ASTM	Color American Standard Test Method
Cole Jct	Coleman Junction
coll	collect (ed) (ing) (ion)
colr	collar
Com	Comanche
Com	Comatula
com	common

Com Pk	Comanche Peak
comb	combined, combination
COMB	combustion
coml	commercial
comm	commenced
comm	commission
comm	communication
comm	community
commr	commissioner
comp	complete (ed) (tion)
comp nat	completed natural
compnts	components
compr	compressor
compr sta	compressor station
compt	compartment
COMPT	component (s)
COMPTR	computer
COMT	comment
COMUT	commutator
con	consolidated
conc	concentrate
conc	concentric
conc	concrete
conch	conchoidal
concl	conclusion
cond	condensate
cond	condition (ed) (ing)
condr	conductor (pipe)
condt	conductivity
conf	confidential
conf	confirm (ed) (ing)
confl	conflict
cong	conglomerate (itic)
conn	connection
cono	conodonts
consol	consolidated
const	constant
const	construction
consv	conserve, conservation
cont (d)	continue (ed)

contam	contaminated, contamination
contr	contractor
contr resp	contractor responsibility
contrib	con⁺ribution
conv	converse
CONVT	convector, convection
COOH	coming out of hole
coord	coordinate
COP	crude oil purchasing
coq	coquina
cor	corner
Corp	corporation
corr	correct (ed) (ion)
corr	corrosion
corr	corrugated
correl	correlation
corres	correspondence
COSH	hyperbolic cosine
COTD	cleaned out to total depth
COTH	hyperbolic cotangent
Cott G	Cottage Grove
Counc G	Council Grove
CO₂	carbon dioxide
CP	casing point
CP	casing pressure
cp	centipoise (s)
cp	chemically pure
CP	correlation point
Cp Colo	Camp Colorado
CPA	certified public accountant
CPC	casing pressure closed
cp'd	cemented through perforations
CPF	casing pressure flowing
CPG	cost per gallon
CPG	cents per gallon
cplg	coupling
CPM	cycles per minute
CPO	confirming telephone order (purchasing term)

CPR	Copper River Meridian (Alaska)
CPS	cycles per second
CPSI	casing pressure shut in
C Riv	Cane River
CR	cold rolled
CR	compression ratio
CR	Cow Run
CR Con	carbon residue (Conradson)
cr moly	chrome molybdenum
cr (d)(g)(h)	core (ed) (ing)
CRA	cased reservoir analysis
CRA	chemically retarded acid
crbd	crossbedded
CRC	Coordinating Research Council Inc.
CRCMF	circumference
crd	cored
CRDL	cradle (s)
cren	crenulated
Cret	Cretaceous
crg	coring
Crin	crinoid (al)
Cris	Cristellaria
crit	critical
crk	creek
crkg	cracking
Crkr	cracker
CRN	crane
crn blk	crown block
crnk	crinkled
Crom	Cromwell
crs	coarse (ly)
CRS	cold-rolled steel
CRS	cross
CRS	retainer
crs-xln	coarse crystalline
CRT	cathode ray tube
CRV	curve
crypto-xln	cryptocrystalline

cryst	crystalline
CS	carbon steel
CS	casing seat
CS	cast steel
cs	centistokes
CSA	casing set at
CSCH	hyperbolic constant
cse gr	coarse grained
csg	casing
csg hd	casinghead
csg press	casing pressure
csg pt	casing point
CSK	countersink
CSL	center section line
CSL	county school lands
CS_2	carbon disulfide
CT	cable tools
CT	cooling tower
CTC	consumer tank car
ctd	coated
CTD	corrected total depth
ctg(s)	cuttings
CTHF	center of tubing flange
Ctlmn	Cattleman
CTM	cable tool measurement
ctn	carton
Ctnwd	Cottonwood
CTP	cleaning to pits
ctr	center
CTS	cement to surface
CTT	consumer transport truck
CTU	coiled tubing unit
CTW	consumer tank wagon
CU	clean up
cu	cubic
cu cm	cubic centimeter
cu ft	cubic foot
cu ft/bbl	cubic feet per barrel
cu ft/min	cubic feet per minute
cu ft/sec	cubic feet per second

cu in.	cubic inch
cu m	cubic meter
cu yd	cubic yard
CUB	cubical
culv	culvert
cum	cumulative
Cur	Curtis
cush	cushion
CUST	customer
Cut B	cut bank
Cut Oil	cutting oil
Cut Oil Act Sul dk	cutting oil active-sulfurized-dark
Cut Oil Act Sul trpt	cutting oil active-sulfurized-transparent
Cut Oil Inact S	cutting oil inactive-sulfurized
Cut Oil Sol	cutting oil soluble
Cut Oil St Mrl	cutting oil straight mineral
cutbk	cutbank
Cutl	Cutler
CV	control valve
CV	Cotton Valley
cvg(s)	cavings (s)
CVO	confirming telephone order (purchasing term)
CVR	cover
CVTR	convert (er) (ed)
cw	clockwise
CW	continuous weld
CW	cooling water
CWE	cold water equivalent
CWP	cold working pressure
CWR	cooling water return
CWS	cooling water supply
cwt	hundredweight
CX	crossover
Cy Sd	Cypress Sand
Cyc	cyclamina
CYC	cyclone

Cycl	cylone
cyl	cylinder
cyn	canyon
Cyp	Cypridopsis
Cz	Carrizo

D

D	day
D	development
D	dual
D & A	dry and abandoned
D & B	Dun & Bradstreet
D & C	drill and complete
D & D	Desk and Derrick
D.I.	diesel index
D.O.	division order
d-d-1-s-1-e	dressed dimension one side and one edge
d-d-4-s	dressed dimension four sides
d-1-s	dressed one side
D-2	Diesel No. 2
d-2-s	dressed two sides
d-4-s	dressed four sides
d/b/a	doing business as
D/D	day to day
D/L	density log
D/O	division office
D/P	differential pressure
D/P	drill (ed) (ing) plug
D/S	data sheet
D/T	driller's top
D/T	drilling tender
DA	daily allowable
DA	Dresser Atlas
DA	drift angle

DAIB	daily average injection barrels
Dak	Dakota
Dan	Dantzler
Dar	Darwin
DAR	discovery allowable requested darcy (darcies not abbreviated)
dat	datum
db	decibel
DB	diamond bit
DB	drilling break
DBA	depth bracket allowable
DBL	double
DBO	dark brown oil
DBOS	dark brown oil stains
DC	delayed coker
DC	development well, carbon dioxide
DC	diamond core
DC	digging cellar or digging cellar and slush pits
DC	direct current
DC	drill collar
DC	dually completed
DCB	diamond core bit
DCLSP	digging slush pits
DCM	distillate-cut mud
DCS	distributed control system
DCTR	detector
dd	dead
DD	degree day
DD	deviation degrees
DD	drilling deeper
DD	dyna-drilling
DDD	dry desiccant dehydrator
DDT	dichloro-diphenyl-trichloroethane
DE	double end
DEA	diethanolamine

DEA unit	diethanolamine unit
Deadw	Deadwood
deaer	deaerator
deasph	deasphalting
debutzr	debutanizer
dec	decimal
decl	decline
decr	decrease (ed) (ing)
deethzr	deethanizer
defl	deflection
Deg	Degonia
deg	degree (s)
deisobut	deisobutanizer
Del R	Del Rio
Dela	Delaware
delv	delivery (ed) (ability)
delv pt	delivery point
demur	demurrage
dend	dendrite (ic)
DENL	density log
depl	depletion
deprec	depreciation
deprop	depropanizer
dept	department
Des Crk	Desert Creek
Des M	Des Moines
desalt	desalter
desc	description
desorb	desorbent
desulf	desulfurizer
det	detail (s)
det	detector
deterg	detergent
detr	detrital
dev	deviate, deviation
Dev	Devonian
devel	develop (ed) (ment)
dewax	dewaxing
Dext	Dexter
DF	derrick floor

DF	diesel fuel
DF	drill floor
DFE	derrick floor elevation
DFO	datum faulted out
DFP	date of first production
DFT	dry film thickness
dg	decigram
DG	development gas well
DG	draft gauge
DG	dry gas
DGA	diglycolamine
DGTL	digital
DH	development well, helium
DH	double hub
DHC	dry-hole contribution
DHDD	dry hole drilled deeper
DHDS	diesel hydrogen desulfurization
DHM	dry-hole money
DHR	dry hole reentered
dia	diameter
diag	diagonal
diag	diagram
diaph	diaphragm
dichlor	dichloride
diethy	diethylene
diff	different (ial) (ence)
DIFL	dual injection focus log
dilut	diluted
dim	dimension
dim	diminish (ing)
Din	Dinwoody
dir	direct (tion) (tor)
dir drlg	directional drilling
dir sur	directional survey
disc	discharge
Disc	Discorbis
disc	discount
disc	discover (y) (ed) (ing)
Disc. grav.	Discorbis gravelli

Disc. norm.	Discorbis normada
Disc. y.	Discorbis yeguaensis
disch	discharge
dism	disseminated
disman	dismantle
displ	displaced, displacement
dist	distance
dist	distillate, distillation
dist	district
distr	distribute (ed) (ing) (ion)
div	division
dk	dark
Dk Crk	Duck Creek
dl	deciliter
DL	drilling line
DLC	dual lower casing
dlr	dealer
DLS	dogleg severity
DLT	dual lower tubing
DM	datum
dm	decimeter
DM	demand meter
DM	dipmeter
DM	drilling mud
dml	demolition
DMPD	dumped
dmpr	damper
DMS	dimethyl sulfide
dm	cubic decimeter
dm/s	cubic decimeter per second
dn	down
dns	dense
DO	development oil
DO	development oil well
do	ditto
DO	drill (ed) (ing) out
DOBLDG	dock operating building
DOC	diesel oil cement
Doc	Dockum
doc	document

DOC	drilled-out cement
doc-tr	doctor-treating
DOCREQ	documentation
DOD	drilled-out depth
DOE	Department of Energy
dolo	dolomite (ic)
dolst	dolstone
dom	domestic
dom AL	domestic airline
DOM WTR	domestic water
DOP	drilled-out plug
Dorn H	Dornick Hills
Doth	Dothan
Doug	Douglas
doz	dozen
DP	data processing
DP	dewpoint
DP	double pipe
DP	drillpipe
DP SW	double pole switch
DPDB	double pole double base (switch)
DPDT SW	double pole double throw switch
dpg	deepening
DPM	drillpipe measurement
dpn	deepen
DPSB	double pole single base (switch)
DPST SW	double pole single throw switch
DPT	deep pool test
dpt	depth
dpt rec	depth recorder
DPU	drillpipe unloaded
DR	development redrill (sidetrack)
dr	drain
dr	drive
dr	drum

dr	druse
Dr Crk	Dry Creek
DRAPR	Delaware River Area Petroleum Refineries
drk	derrick
DRL	double random lengths
drl	drill
drld	drilled
drlg	drilling
drlr	driller
DRM	drum
drng	drainage
dropd	dropped
drsy	drusy
DRV (R)	drive (ing) (er)
dry	drier, drying
ds	dense
DS	directional survey
DS	drillsite
DS	drillstem
DSF	drillsite facility
dsgn	design
DSI	drilling suspended indefinitely
dsl	diesel (oil)
dsmt (g)	dismantle (ing)
DSO	dead oil show
DSS	days since spudded
DST	drillstem test
DST (Strd)	drillstem test with straddle packers
dstl	distillate
dstn	destination
DSU	development well, sulfur
DSUPHTR	desuperheater
DT	downthrown
DT	drilling time
DTD	driller's total depth
dtr	detrital
DTW	dealer tank wagon

DUC	dual upper casing
Dup	Duperow
dup	duplicate
DUR	duration
DUT	dual upper tubing
Dutch	Dutcher
DV	differential valve (cementing)
DVL	develop
DVT	davit
DWA	drilling with air
DWC	drilling and well completion
DWD	dirty water disposal
DWG	drawing
DWG	drilling with gas
dwks	drawworks
DWM	drilling with mud
DWO	drilling with oil
DWP	dual (double) wall packer
DWSW	drilling with salt water
DWT	deadweight tester
DWT	deadweight tons
DWTR	dewatering
DX	development well workover
dx	duplex
dyn	dynamic

E

E	east
E	exploratory
E of W/L	east of west line
E.D.	effective depth
e.g.	for example
E.T.D.	estimated total depth
E/BL	east boundary line
E/E	end to end

E/L	east line
E/O	east offset
E/2	east half
E/4	east quarter
ea	each
EA	environmental assessment
EAM	electric accounting machines
Earls	Earlsboro
Eau Clr	Eau Claire
ECC	eccentric
Ech	echinoid
ECM	East Cimarron Meridian (Oklahoma)
Econ	economics, economy, economizer
Ect	Ector (County, TX)
ecx	excavation
Ed lm	Edwards lime
EDC	ethylene dichloride
EDCTN(R)	education (tor)
EDD	expected date of delivery
EDP	electronic data processing
Educ	education
Edw	Edwards
EF	Eagle Ford
eff	effective
eff	efficiency
effl	effluent
EFV	equilibrium flash vaporization
Egl	Eagle
Eglwd	Englewood
EHP	effective horsepower
EIA	environmental assessment
EIR	environmental impact report
EIS	environmental impact statement
EJ	perforating, enerjet

eject	ejector
EL	elevation (height)
el gr	elevation ground
EL/T	electric log tops
Elb	elbert
ELB	elbow
elec	electric (al)
Elec/MAG	electromagnetic
elem	element, elementary
elev	elevation, elevator
Elg	Elgin
ELIM	eliminate (tor) (ed)
ell(s)	elbow (s)
Ellen	Ellenburger
Elm	Elmont
E'ly	easterly
EM	Eagle Mills
Emb	Embar
emer	emergency
EMF	electromotive force
EMN	electromagnetic
EMNI	electromagnetic induction
EMP	European melting point
empl	employee
EMS	Ellis-Madison contact
emul	emulsion
encl	enclosure
End	Endicott
endo	endothyra
eng	engine
engr (g)	engineer (ing)
enl	enlarged
enml	enamel
Ent	Entrada
ENT	entrance
ent	entry
ENV	envelope
ENVIR	environment
EO	emergency order
Eoc	Eocene

EOF	end of file
EOL	end of line
EOM	end of month
EOQ	end of quarter
EOR	east of Rockies
EOR	ehhanced oil recovery
EOY	end of year
EP	end point
EP	extreme pressure
Epon.	Eponides
Epon. y.	Eponides yeguaensis
eq	equal, equalizer
Eq	equation (before a number)
equip	equipment
equiv	equivalent
ERC/MK	erection mark
erect	erection
Eric	Ericson
ERW	electric resistance weld
ESD	emergency shutdown
est	estate
est	estimate (ed) (ing)
et al.	and others
et con.	and husband
et seq.	and the following
et ux.	and wife
et vir.	and husband
ETA	estimated time of arrival
eth	ethane
ethyle	ethylene
EU	Eutaw
EUE	external upset end
euhed	euhedral
EUR	estimated ultimate recovery
ev	electron-volts
ev-sort	even sorted
eval	evaluate
evap	evaporation, evaporate
EW	electric weld
EW	exploratory well

EWT	early well tie-ins
EX	example
ex	except
Ex	Exeter
EX-PRF	explosion proof
EXAM	examination
EXC	excitation
exch	exchanger
excl	excellent
EXEC	executive
exh	exhaust
exh	exhibit
exist	existing
EXP	expansion
exp	expense
EXP JT	expansion joint
exp plg	expendable plub
expir	expire (ed) (ing) (ation)
expl	exploratory, exploration
explos	explosive
exr	executor
Exrx	executrix
exst	existing
ext	external
Ext M/H	extension manhole
ext(n)	extended, extension
extr	exterior
extrac	extraction
EYC	estimated yearly consumption

——————— **F** ———————

F & D	faced and drilled
F & D	flanged and dished (heads)
F & F	fuels & fractionation
F & L	fuels & lubricants
F & S	flanced and spigot

F to F	face to face
F.G.	fracture gradient
F.O.E.	fuel oil equivalent
F-D	formation density
F-DIA	flow diagram
f-gr	fine grained
F-MET	flowmeter
F-R oil	fire-resistant oil
F-SHT	flow sheet
F/	flowed, flowing
F/DIA	flow diagram
F/FAB	field fabricated
F/GOR	formation gas-oil ratio
F/O opt	farmout option
F/S	flange & screwed
F/S	front & side
F/SW	flow switch
F/WTR	fire water
f/xln	finely crystalline
fab	fabricate (ed) (tion)
FAB	faint air blow
fac	facet (ed)
FACIL	facility (ies)
FACO	field authorized to commence operations
fail	failure
Fall Riv	Fall River
FAO	finish all over
Farm	Farmington
FARO	flowed (ing) at a rate of
fau	fauna
FB	fresh break
FBH	flowing by heads
FBHP	flowing bottom-hole pressure
FBHPF	final bottom-hole pressure, flowing
FBHPSI	final bottom-hole pressure, shut-in
FBP	final boiling point

FC	filter cake
FC	fixed carbon
FC	float collar
FCC	fluid catalytic cracking
FCL	facility capacity limits
FCO	functional check out
FCP	flowing casing pressure
FCV	flow control valve
FD	feed
FD	floor drain
FD	flow diagram
FD EFF	feed effluent
FD/WTR	feed water
FDC	formation density correlated
FDL	formation density log
fdn	foundation
fdr	feeder
Fe	iron
Fe-st	ironstone
FE/L	from east line
fed	federal
FEIS	Final Environmental Impact Statement
FEL	from east line
FELA	Federal Employers Liability Act
FEM	female
Ferg	Ferguson
ferr	ferruginous
fert	fertilizer
$Fe_2(SO_4)_3$	ferric sulfate
FF	fishing for
FF	flat face
FF	frac finder (log)
FF	full of fluid
FFA	female to female angle
FFA	full freight allowed (purchasing term)
FFG	female to female globe (valve)

FFGU	field fuel gas unit
FFL	final fluid level
FFO	furnace fuel oil
FFP	final flowing pressure
FG	fuel gas
FGIH	finish going in hole
FGIW	finish going in with
FGVV	flanged gate valve
FH	full hole
FHP	final hydrostatic pressure
FI	flow indicator
fib	fibrous
FIC	flow-indicating controller
fig	figure
FIH	finished in hole
FIH	fluid in hole
filt	filtrate
fin	final
fin	finish (ed)
fin drlg	finished drilling
FIN GR	finish grade
FIRC	flow indicating ratio controller
fis	fissure
fish	fishing
fisl	fissile
FIT	formation interval tester
fix	fixture
FJ	flush joint
fkt	fault
FL	flashing
FL	floor
FL	flow line
fl	fluid
FL	fluid level
FL	flush
Fl-COC	flash point, Cleveland Open Cup
fl/	flowed (ing)
FL/BD	flammable liquid building

FLA	Ferry Lake anhydrite
flat	flattened
Flath	Flathead
fld	failed
fld	feldspar (thic)
fld	field
FLD	full-length drift
flex	flexible
flg	flowing
flg (d) (s)	flange (ed) (es)
Flip	Flippen
flk	flaky
FLMB	flammable
flo	flow
FLO	flushing oil
Flor fl	Florence flint
flshd	flushed
flt	float
fltg	floating
fltn	flotation
FLTR	filter
flu	flue
flu	fluid
fluor	fluorescence, fluorescent
flw (d) (g)	flow (ed) (ing)
Flwg Pr.	flowing pressure
Flwrpt	flowerpot
FLXBX	flexibox
fm	formation
FM	frequency meter
FM	frequency modulation
Fm W	formation water
f'man	foreman
fmp	feet per minute
fn	fine
FNEL	from northeast line
FNL	from north line
fnly	finely
FNSH	finish
fnt	faint

FNWL	from northwest line
FO	farmout
FO	faulted out
FO	final open
FO	fuel oil
FO	full opening
FOB	free on board
FOCL	focused log
FOE-WOE	flanged one end, welded one end
FOH	full open head
fol	foliated
FONSI	finding of no significant impact
FOR	fuel oil return
Forak	Foraker
foram	foraminifera
Fort	Fortura
FOS	face of stud
FOS	fuel oil supply
foss	fossiliferous
FOT	flowing on test
Fount	Fountain
Fox H	Fox Hills
FP	final pressure
FP	flowing pressure
FP	freezing point
FPI	free-point indicator
FPO	field purchase order
fprf	fireproof
fps	feet per second
f-p-s	foot-pound-second (system)
FPT	female pipe thread
FPTFD	field pressure test flow diagram
FQG	frosted quartz grains
fr	fair
FR	feed rate
FR	flow rate
FR	flow recorder

fr	fractional
fr	from
fr	front
fr	frosted
fr E/L	from east line
fr N/L	from north line
fr S/L	from south line
fr W/L	from west line
FRA	friction reducing agent
frac (d) (s)	fracture, fractured, fractures
fract	fractionation, fractionator, fractional
frag	fragment
fran	franchise
Franc	Franconia
FRC	flow recorder control
Fred	Fredericksburg
Fred	Fredonia
freq	frequency
FRG	forge (ed) (ing)
Frgy	froggy
fri	friable
FRM (G)	frame, framing
Fron	frontier
fros	frosted
FRP	fiberglass-reinforced plastic
FRR	field receiving report
FRR	final report for rig
frs	fresh
frt	freight
Fruit	Fruitland
FRW	final report for well
frwk	framework
frzr	freezer
FS	feedstock
FS	float shoe
FS	flow station
FS	forged steel
FS&WLs	from south and west lines

FSEL	from southeast line
fsg	fishing
FSIP	final shutin pressure
FSIWA	Federation of Sewage and Industrial Wastes Association
FSL	from south line
FSP	flowing surface pressure
FST	forged steel
FSTN	fasten (ing) (er)
FSWL	from southwest line
ft	feet, foot
FT	formation test
Ft C	Fort Chadborne
Ft H	Fort Hayes
ft-lb	foot-pound
ft-lb/hr	foot-pound per hour
Ft R	Fort Riley
Ft U	Fort Union
Ft W	Fort Worth
ft-c	foot-candle
ft/hr	feet per hour
ft/min	feet per minute
ft/sec	feet per second
ftg	fittings
ftg	footing, footage
FTP	final (flowing) tubing pressure
FTS	fluid to surface
ft²	square feet
ft³	cubic feet
FU	fill up
Full	Fullerton
furf	furfural
furn	furnance
FURN	furnish (ed)
Furn & fix	furniture and fixtures
Fus	Fuson
Fussel	Fusselman
Fusul	Fusulinid

fut	future
FV	funnel viscosity
fvst	favosites
FW	fillet weld
FW	fresh water
FWC	field wildcat
fwd	forward
FWD	four-wheel drive
FWL	from west line
fwtr	fresh water
fxd	fixed
FYE	fiscal year ending
FYI	for your information

G

G	gas
g	gram
G egg	goose egg
g mole	gram molecular weight
G Rk	gas rock
G.M.	gravity meter
G. Riv	Gull River
g-cal	gram-calorie
G-N$_2$	gaseous nitrogen
G-O$_2$	gaseous oxygen
G&MCO	gas & mud-cut oil
G&O	gas and oil
G&OCM	gas and oil-cut mud
G/L	gathering line
G/O	gas-oil ratio
G/P	gun perforate
ga	gauge (ed) (ing)
GA	gallons acid
GA	general arrangement
GAF	gross acre-feet
gal	gallon (s)

gal sol	gallons of solution
gal/Mcf	gallons per thousand cubic feet
gal/min	gallons per minute
Gall	Gallatin
galv	galvanized
gaso	gasoline
gast	gastropod
GB	gun barrel
GBDA	gallons breakdown acid
GC	gas-cut
GCAW	gas-cut acid water
GCD	gas-cut distillate
GCLO	gas-cut load oil
GCLW	gas-cut load water
GCM	gas-cut mud
GCO	gas-cut oil
GCPA	gas cap participating area
GCPD	gallons condensate per day
GCPH	gallons condensate per hour
GCR	gas-condensate ratio
GCSW	gas-cut salt water
GCT	guidance continuance tool
GCW	gas-cut water
GD	geothermal development, failure
GD li	Glen Dean lime
gd	good
gd o&t	good odor & taste
Gdld	Goodland
GDR	gas-distillate ratio
GDS	geothermal development, success
Gdwn	Goodwin
GE	General Electric Company
GE	grooved ends
gel	gelled
gel	jelly-like colloidal suspension
gen	generation, generator

genl	general
Geo	Georgetown
geo	geothermal
geol	geology (ist) (ical)
geop	geophysics (ical)
GFLU	good fluorescence
GFR	gas-fluid ratio
gg	grains per gallon
GGD	gas lift gas distribution
gge	gauge
GGW	gallons gelled water
GH	Greenhorn
GHO	gallons heavy oil
GHSG	gas-handling study group
GI	gas injection
Gib	Gibson
GIH	going in hole
gil	gilsonite
Gilc	Gilcrease
GIP	gas in pipe
GIW	gas injection well
GL	gas lift
GL	ground level
glau	glauconite, glauconitic
GLBVV	globe valve
gld thd	galled threads
Glen	Glenwood
Glna	Galena
GLO	gas lift oil
GLO	General Land Office (Texas)
Glob	Globigerina
Glor	Glorieta
GLR	gas-liquid ratio
gls (y)	glass, glassy
GLT	gas lift transfer
glyc	glycol
GM	General Motors Corporation
gm	gram
GM	ground measurement (elevation)

gm-cal	gram-calorie
GMA	gallons mud acid
gmy	gummy
gnd	grained (as in fine-grained)
gns	gneiss
GO	gallons oil
GO	gas odor
GO	grind out
GOC	gas-oil contact
GODT	gas odor distillate taste
Gol	Golconda lime
Good L	Goodland
GOPD	gallons of oil per day
GOPH	gallons of oil per hour
GOR	gas-oil ratio
Gor	Gorham
Gouldb	Gouldbusk
gov	governor
govt	government
GP	gas pay
GP	gasoline plant
GPC	gas purchase contract
GPD	gallons per day
GPF	granite point field
GPG	grains per gallon
GPH	gallons per hour
GPM	gallons per minute
GPM	geophysical investigation map
GPS	gallons per second
GQM	geological quadrangle map
GR	gamma ray
GR	gauge ring
GR	Glen Rose
gr	grade
gr	grain
gr	gravity
gr	grease
gr	ground
gr API	gravity °API

gr roy	green royalty
Gr Sd	Gray sand
gr wt	gross weight
GR&DC	Gulf Research and Development Company
GRA	gallons regular acid
GRAD	gradiomanometer
grad	gradual, gradually
gran	granite, granular
gran w	granite wash
Granos	Graneros
grap	graptolite
grav	gravity
Gray	Grayson
Grayb	Grayburg
grdg	grading
grdg loc	grading location
GRDL	guard log
GRG	gas reserve group
grn	green
Grn Riv	Green River
grn sh	green shale
grnd	ground
grnlr	granular
GRP	group
GRS	gas to surface
grs	gross
grt	grant (of land)
grtg	grating
grty	gritty
grv	grooved
grvt	gravitometer
GRVTY	gravity
gry	gray
GS	gas show
GS	guide shoe
GSC	gas sales contract
GSG	good show of gas
GSI	gas well shutin
gskt	gasket

GSO	good show oil
GSO&G	good show oil and gas
GST	gamma spectroscopy tool
GSW	gallons salt water
gsy	greasy
GT	geothermal
GTS	gas to surface (time)
GTSTM	gas too small to measure
GTY	gravity
GU	gas unit
Guns	Gunsite
GUS	gusset
GV	gas volume
gvl	gravel
GVLPK	gravel packed
GVNM	gas volume not measured
GW	gallons water
GW	gas well
GW	geothermal wildcat, failure
GWC	gas-water contact
GWD	geothermal wildcat, success
GWG	gas-well gas
GWPH	gallons of water per hour
gyp	gypsum
gypy	gypsiferous
Gyr.	Gyroidina
Gyr. sc.	Gyroidina scal
gywk	graywacke

--------- **H** ---------

H & V	heating and ventilating
H H P	hyraulic horsepower
H. O.	hole opener
H-SEL	perforating hyper select
H-VOLT	high voltage
H/C	hydrocracker
Hackb	Hackberry

Hara	Haragan
Hask	Haskell
Haynes	Haynesville
haz	hazardous
HB	house brand (regular grade of gasoline)
Hburg	Hardinsburg sand (local)
HBP	held by production
hbr	harbor
HC	hydrocarbon
HC	hydrocracker
HCDS	hydrocarbon drain system
HCGO	heavy coker gas oil
HCO	heavy cycle oil
HCV	hand-control valve
hd	hard
hd	head
HD	heavy duty
HD	high detergent
HD	hot dry rock development, failure
HD	Hydril
HD	perforating, hyperdome
Hd li	hard lime
hd sd	hard sand
hdl	handle
hdns	hardness
hdr	header
HDS	hot dry rock development, successful
HDS	hydrogen delsulfurization
hdwe	hardware
Heeb	Heebner
hem	hematite
Her	Herington
Herm	Hermosa
het	heterostegina
HEX	heat exchanger
hex	hexagon (al)
hex	hexane

HEX HD	hex head
hfg	hydrofining
HFO	heavy fuel oil
HFO	hole full of oil
HFSW	hole full of salt water
HFW	hole full of water
HGCM	heavily (highly) gas-cut mud
HGCSW	heavily (highly) gas-cut salt water
HGCW	heavily (highly) gas-cut water
HGOR	high gas-oil ratio
hgr	hanger
hgt	height
HH	hand hole
HH	hydrostatic head
HIA	Hydrologic Investigations Atlas
Hick	Hickory
Hill	Hilliard
hky	hackly
HLDN	hold down
HLSD	high-level shutdown
HND/WHL	handwheel
HNDLG	handling
HO	heating oil
HO	heavy oil
HO	hole opener
HO	home office
HO&GCM	heavily (highly) oil- and gas-cut mud
hock	hockleyensis
HOCM	heavily (highly) oil-cut mud
HOCSW	heavily (highly) oil-cut salt water
HOCW	heavily (highly) oil-cut water
Hog	Hogshooter
Holl	Hollandberg

Home Cr	Home Creek
hop	hopper
horiz	horizontal
Hosp	Hospah
HOT	hot oil tar
Hov	Hoover
Hox	Hoxbar
HP	high pressure
HP	horsepower
HP	hydraulic pump
HP	hydrostatic pressure
hp-hr	horsepower-hour
HPF	holes per foot
HPG	high-pressure gas
HPG	high-pressure gauge
HQ	headquarters
HR	heavy reformate
hr	hour (s)
HR Sul W	hole full of sulfur water
HRD	high-resolution dipmeter
hrs	heirs
HRS	hot-rolled steel
HSD	heavy steel drum
HSE	house (ed) (ing)
HST	hydrostatic test
HT	heat tracing (ed)
HT	heat-treated, heater treater
HT	high temperature
HT	high tension
HTA	heat-treated alloy
htr	heater
HTSD	high-temperature shutdown
HU	hook up
Humb	Humblei
Hump	Humphreys
Hun	Hunton
HUX	heavy hydrocrackate
HV	high viscosity
HVAC	heating ventilating and air conditioning

HVGO	heavy gas oil
HVI	high viscosity index
HVL	high volume lift
hvly	heavily
hvy	heavy
HW	hot dry rock wildcat, failure
HWCM	heavily (highly) water-cut mud
HWD	hot dry rock wildcat, success
HWP	hookwall packer
hwy	highway
HX	heat exchanger
HYD	hydraulic
HYD	Hydril thread
HYDA	Hydril Type A joint
HYDCA	Hydril Type CA joint
HYDCS	Hydril Type CS joint
HYDRO	hydro test
hydtr	hydrotreater
Hyg	hygiene
HYGN	hydrogenation
HYPO	hypotenuse
Hz	Hertz
H$_2$	hydrogen
H$_2$S	hydrogen sulfide
H$_2$SO$_4$	sulfuric acid

I

I.D. sign	identification sign
I-	miscellaneous investigations series
I-O-M	installation operation & maintenance
I/C	interconnection (ing)

I/CFD	interconnecting flow diagram
I/O	input/output
IAB	initial air blow
IB	impression block
IB	iron body (valve)
IBBC	iron body brass core (valve)
IBBM	iron body brass (bronze) mounted (valve)
IBHP	initial bottom-hole pressure
IBHPF	initial bottom-hole pressure, flowing
IBHPSI	initial bottom-hole pressure, shut in
IBP	initial boiling point
IC	iron case
icfos	microfossil(iferous)
ID	inside diameter
IDENT	identify (ier) (ication)
Idio	Idiomorpha
IES	induction electrical survey
IF	internal flush
IFL	initial fluid level
IFP	initial flowing pressure
IG	injection gas
Ign	igneous
IGN	ignition
IGOR	injection gas-oil ratio
IGV/IBV	inlet gate valve/inlet ball valve
IH	in hole
IHP	indicated horsepower
IHP	initial hydrostatic pressure
IHPHR	indicated horsepower hour
II	injection index
IJ	integral joint
ILUM	illuminator (s)
imbd	imbedded
IMF	intermediate manifolds
immed	immediate (ly)

Imp	Imperial
IMP	impounding
Imp gal	Imperial gallon
imperv	impervious
IMW	initial mud weight
in.	inch (es)
in. Hg	inches mercury
in.-lb	inch-pound
in./sec	inches per second
Inbded	inbedded
Inc.	incorporated
inc	increment
incd	incandescent
INCIN	incinerator, incineration
incl	include (ed) (ing)
INCLR	intercooler
incls	inclusions
INCM	income (er) (ing)
incolr	intercooler
incr	increase (ed) (ing)
ind	induction
indic	indicate (s) (tion)
indiv	individual
indr	indurated
indst	indistinct
Inf. L	inflammable liquid
Inf. S	inflammable solid
info	information
ingr	intergranular
inhib	inhibitor
init	initial
inj	injection, injected
Inj Pr	injection pressure
inl	inland
inl	inlet
inlam	interlaminated
Inoc	Inoceramus
INPE	installing (ed) pumping equipment
INQ	inquire, inquiry

ins	insulate, insulation
ins	insurance
insol	insoluble
insp	inspect (ed) (ing) (tion)
inst(d)(g)(l)	install (ed) (ing) (ation)
inst	instantaneous
inst	institute
instr	instrument, instrumentation
insul	insulate
INT	integral
int	interest
int	interior
int	internal
int	intersection
INTCON	interconnection
inter-gran	intergranular
inter-lam	interlaminated
inter-xln	intercrystalline
interbd	interbedded
intgr	integrator
INTL	internal
intl	interstitial
intr	intrusion
ints	intersect
intv	interval
inv	invert (ed)
inv	invoice
invrtb	invertebrate
IO	initial open
IP	initial potential
IP	initial pressure
IP	initial production
IP	intermediate pressure
IPA	initial participating area
IPA	isopropyl alcohol
IPE	install (ing) pumping equipment
IPF	initial production flowed (ing)

IPG	initial production gas lift
IPI	initial production on intermitter
IPL	initial production plunger lift
IPP	initial production pumping
IPR	inflow performance rate
IPS	initial production swabbing
IPS	iron pipe size
IPT	iron pipe thread
IR	infrared
IR	injection rate
Ire	Ireton
irid	iridescent
irreg	irregular
IRS	Internal Revenue Service
irst	ironstone
IS	inside screw (valve)
ISIP	initial shutin pressure (DST)
ISIP	instantaneous shutin pressure (frac)
ISITP	initial shutin tubing pressure
ISO	isometric
ISO/CKR	isocracker
ISOL	isolate (tor)
isom	isometric
isoth	isothermal
ISS	issue
ITB	invitation to bid
ITC	investment tax credit
ITD	intention to drill
IUE	internal upset ends
Ives	Iverson
IVP	initial vapor pressure
IW	injection water
IW	injection well

—————— **J** ——————

J	Joule
J&A	junked and abandoned
J/O	joint operation
jac	jacket
Jack	Jackson
Jasp	jasper (oid)
Jax	Jackson sand
JB	junction box
JB	junk basket
jbr	jobber
JC	job complete
jct	junction
JCUMWE	Joint Committee on Uniformity of Methods of Water Examination
Jdn	Jordan
Jeff	Jefferson
JFA	jet fuel (aviation)
JINO	joint interest nonoperated (property)
JJ	junk joint
jmd	jammed
jnk	junk (ed)
JOA	joint operating agreement
JOP	joint operating provisions
JP	jet perforated
JP fuel	jet propulsion fuel
JP/ft	jet perforations per foot
JSPF	jet shots per foot
jt(s)	joint (s)
JTU	jet treating unit
Jud Riv	Judith River
Jur	Jurassic
juris	jurisdiction
JV	joint venture
Jxn	Jackson

K

K	Kelvin (temperature scale)
K	thousand (i.e., 13K = 13,000)
Kai	Kaibab
kao	kaolin
Kay	Kayenta
KB	kelly bushing
KBM	kelly bushing measurement
KC	Kansas City
kc	kilocycle
kcal	kilocalorie
KD	kiln dried
KD	Kincaid lime
KD	knock down
KDB	kelly drive bushing
KDB-LDG FLG	kelly drill bushing to landing flange
KDB-MLW	kelly drill bushing to mean low water
KDB-Plat	kelly drill bushing to platform
KDBE	kelly drive bushing elevation
Ke	Keener
Keo-Bur	Keokuk-Burlington
kero	kerosine
ket	ketone
kev	thousand electron-volts
Key	Keystone
kg	kilogram
kg-cal	kilogram-calorie
kg-m	kilogram-meter
KGRA	known geothermal resource area
Khk	Kinderhook
kHz	kilohertz
Kia	Kiamichi
Kib	Kibbey

Kin li	Kincaid lime
kin	kinematic
kip	one thousand pounds
kip-ft	one thousand foot-pounds
Kirt	Kirtland
KIT	kitchen
kl	kiloliter
kld	killed
km	kilometer
KMA	KMA sand
KO	kick off
KO	knock out
Koot	Kootenai
KOP	kickoff point
kPa	kilopascal
Kri	Krider
KTLE	kettle
kv	kilovolt
KV	kinematic viscosity
KV	permeability (vertical direction)
kva	kilovolt-ampere
kvah	kilovolt-ampere-hour
kvar	kilovar; reactive kilovolt-ampere
kvar-hr	kilovar-hour
kvp	kilovolt peak
KW	kill (ed) well
kw	kilowatt
kwh	kilowatt-hour
kwhm	kilowatt-hour meter

L

l	liter
L & P	ladder & platform
L U	lease use (gas)
L.P.	line pipe

L-DK	loading dock
L-VOLT	low voltage
L/	Lower, i.e., L/Gallup
L/Alb	Lower Albany
L/Cret	Lower Cretaceous
L/Tus	Lower Tuscaloosa
LA	level alarm
LA	lightning arrester
LA	load acid
La Mte	La Motte
lab	labor
lab	laboratory
LACT	lease automatic custody transfer
lad	ladder
LAG	lagging
Lak	Lakota
lam	laminated, lamination (s)
Land	Landulina
Lans	Lansing
Lar	Laramie
LAS	lower anhydrite stringer
lat	latitude
Laud	Lauders
Layt	Layton
LB	light barrel
lb	pound
lb-in.	pound-inch
lb/ft	pounds per foot
lb/sq ft	pounds per square foot
LBOS	light brown oil stain
lbr	lumber
LC	lease crude
LC	level controller
LC	long coupling
LC	lost circulation
LC	lower casing
LC	lug cover type (5-gallon can)
LCGO	light coker gas oil
lchd	leached

LCL	less-than-carload lot
LCL	local
LCM	lost circulation material
LCP	local control panel
LCP	lug cover with pour spout
LCV	level control valve
LD	laid down
ld	load
ld(s)	land (s)
LDC	laid-down cost
LDCX	lead drill collar
LDDCs	laid (laying) down drill collars
LDDP	laid (laying) down drillpipe
LDF	large-diameter flow line
LDG	landing
LDG	loading
LDR	loader
Le C	Le Comptom
Leadv	Leadville
LEL	lower explosive limit
Len	Lennep
len	lenticular
LFO	light fuel oil
lg	large
lg	length
lg	level glass
lg	long
Lg Disc	Large Discorbis
LGD	Lower Glen Dean
Lge	league
LH	left hand
LH/RP	long handle/round point
LI	level indicator
li	lime, limestone
LIB	light iron barrel
LIC	level indicator controller
lic	license
Lieb	Liebuscella
lig	lignite, lignitic

LIGB	light iron grease barrel
LIH	left in hole
lim	limit, limonite
lin	linear
lin	liner
lin ft	linear foot
LIP	local injection plants
liq	liquid
liqftn	liquefaction
litho	lithographic
LJ	lap joint
lk	leak
lk	lock
LKG	leakage
LKR	locker
LLC	liquid level controller
LLG	liquid level gauge
lm	lime, limestone
LMF	lowermost flange
LMn	Lower Menard
Lmpy	lumpy
LMTD	log mean temperature difference
lmy	limy
Lmy sh	limy shale
LN	line
ln	logarithm (natural)
LNG	liquefied natural gas
lngl	linguloid
lnr	liner
lns	lense
LO	load oil
LO	lube oil
LOA	length overall
loc	located, location
loc abnd	location abandoned
loc gr	location graded
log	logarithm (common)
long	longitude (inal)
LOS	lease operations

Lov	Lovell
Lov	Lovington
low	lower
LOX	liquid oxygen
LP	line pressure
LP	lodge pole
lp	loop
LP	low pressure
LP sep	low-pressure separator
LP-Gas	liquefied petroleum gas
LPG	liquefied petroleum gas
LPG	propane
LPO	local purchase order
LPS	low-pressure separation
LR	level recorder
LR	long radius
LRAP	long-range automation plan
LRC	level recorder controller
lrg	large
LRP	long-range plan
ls	limestone
LS	long string
LSD	legal subdivision (Canada)
LSD	light steel drum
lse	lease
LST/COMPTS	list of components
lstr	lustre
lt	light
LT	lower tubing
LT&C	long threads and coupling
ltd	limited
LTD	log total depth
ltg	lighting
LTL	less than truckload
ltl	little
ltr	letter
LTS unit	low-temperature separation unit
LTSD	low-temperature shutdown
LTT	long-term tubing test

LTX unit	low-temperature extraction unit
lub	lubricate (ed) (ing) (tion)
Lued	Lueders
LUX	light hydrocrackate
LV	liquid volume
LVI	low viscosity index
lvl	level
Lvnwth	Leavenworth
LW	lapweld
LW	load water
LWL	low water loss
lwr	lower

M

M	mesh
m	meter
m	micron
m	millimicron
M	molar
M	thousand (i.e., 9M = 9,000)
M & R Sta.	measuring and regulating station
M. Tus	Marine Tuscaloosa
m-a	microampere
m-f	microfarad
m-g	microgram
m-gr	medium grained
m-in.	microinch
m-kg	meter-kilogram
m-m	micromicron
m-mf	micro-microfarad
m-v	microvolt
M&F	male and female (joint)
M&FP	maximum & final pressure
M&P	mix and pump
M/	middle

m/l	more or less
M/PLT	masking plate
M/T	marine terminal
M/V	motor vehicle, motor vessel
MA	massive anhydrite
ma	microampere
ma	milliampere
MA	mud acid
MAC	medium amber cut
mach	machine
Mack	Mackhank
Mad	Madison
mag	magnetic, magnetometer
maint	maintenance
maj	major, majority
mall	malleable
man	manifold
man	manual
man op	manually operated
Manit	Manitoban
Mann	Manning
MAOP	maximum allowable operating pressure
Maq	Maquoketa
mar	marine
mar	maroon
March	Marchand
marg	marginal
Marg.	Marginulina
Marg. coco.	Marginulina coco
Marg. fl.	Marginulina flat
Marg. rd	Marginulina round
Marg. tex.	Marginulina texana
margas	marine gasoline
Mark	Markham
Marm	Marmaton
MARSH	Marshal (ling)
mass	massive
Mass. pr.	Massilina pratti
Mat	matter

Math	mathematics
matl	material
MAW	mud acid wash
MAWP	maximum allowable working pressure
max	maximum
May	Maywood
MB	methylene blue
MB	Moody's Branch
MBF/D	thousand barrels fluid per day
MBFPD	thousand barrels fluid per day
Mbl Fls	Marble Falls
Mbo/d	thousand barrels oil per day
mbr	member (geologic)
MBTU	thousand British thermal units
MBW/D	thousand barrels of water per day
mc	megacycle
MC	mud cake
MC	mud cut
MC ls	Moore County lime
MCA	mud cleanout agent
MCA	mud-cut acid
MCB	master circuit board
McC	McClosky lime
MCC	motor control center
McCul	McCullough
McEl	McElroy
Mcfd	thousand cubic feet per day
Mcfgpd	thousand cubic feet of gas per day
MCG	mud-cut gas
mchsm	mechanism
McK	McKee
McL	McLish
McMill	McMillan
MCO	mud-cut oil

MCP	maxium casing pressure
mcr-x	microcrystalline
MCSW	mud-cut salt water
MCT	computer-processed interpretation
MCW	mud-cut water
MD	measured depth
md	millidarcies
MD	Mt. Diablo
md wt	mud weight
MDC	Monel drill collars
MDDO	maximum daily delivery obligation
MDF	market demand factor
mdl	middle
mdse	merchandise
Mdy	muddy
MEA	monoethanolamine
Meak	Meakin
meas	measure (ed) (ment)
mech	mechanic (al), mechanism
Mech DT	mechanical down time
med	median
Med	Medina
med	medium
Med B	Medicine Bow
med FO	medium fuel oil
med gr	medium grained
Medr	Medrano
Meet	Meeteetse
MEG	methane-rich gas
MEK	methylethylketone
memo	memorandum
Men	Menard lime
Mene	Menefee
MEOH	methanol
MEP	mean effective pressure
MER	maximum efficient rate
Mer	Meramec
merc	mercury

mercap	mercaptan
merid	meridian
Meso	Mesozoic
meta	metamorphic
meth	methane
meth-bl	methylene blue
meth-cl	methyl chloride
methol	methanol
methr	methanator
metr	metric
mev	million electron volts
mezz	mezzanine
MF	manifold
MF	mud filtrate
MF-	Miscellaneous Field Studies Map
MFA	male to female angle
mfd	manufactured
MFD	mechanical flow diagram
mfd	microfarad
mfg	manufacturing
MFP	maximum flowing pressure
MFR	manufacture (er)
mg	medium grained
mg	milligram
mg	motor generator
MG	multigrade
MG	thousand gallons
m'gmt	management
mgr	manager
MGS	middle ground shoals
MH	manhole
mh	millihenry
MH	mousehole
mho/m	mhos per meter
MHz	megahertz (megacycles per second)
MI	malleable iron
MI	mile (s)
MI	mineral interest

MI	moving in (equipment)
mica	mica, micaceous
MICR	moving in completion rig
micro-xln	microcrystalline
microsec	microsecond
MICT	moving in cable tools
MICU	moving in completion unit
mid	middle
Mid	Midway
MIDDU	moving (moved) in double drum unit
MIE	moving in equipment
MIK	methylisobutylketone
mil	military
mill	milliotitic
millg	milling
MIM	moving in materials
min	minerals
min	minimium
min	minute(s)
min P	minimum pressure
Minl	Minnelusa
Mio	Miocene
MIPU	moving in pulling unit
MIR	moving in rig
MIRT	moving in rotary tools
MIRU	moving in and rigging up
MIRUSU	moving in rigging up swabbing unit
misc	miscellaneous
Mise	Misener
MISR	moving in service rig
Miss	Mississippian
Miss Cany	Mission Canyon
MIST	moving in standard tools
MIT	moving in tools
MIU	moisture impurities and unsaponifiables (grease testing)
mix	mixer

MIXG	mixing
MKG	making
mkt	market (ing)
Mkta	Minnekahta
mky	milky
ml	milliliter
ML	mud logger
ml TEL	milliliters tetraethyl lead per gallon
mld	milled
MlE	milled one end
mlg	milling
MLL	master load list
MLU	mud logging unit
MLW	mean low wave
MLW-PLAT	mean low water to platform
Mly	marly
mm	millimeter
MM	million (i.e., 9MM = 9,000,000)
MM	motor medium
mm Hg	millimeters of mercury
MMBLS	millions of barrels
MMBTU	million British thermal units
MMcf	million cubic feet
MMcfd	million cubic feet per day
mmf	magnetomotive force
MMRVB	million reservoir barrels
MMS	Minerals Management Service
MMscfd	million standard cubic feet per day
MNL	manual
MNR	minor
mnrl	mineral
MO	molybdenum
MO	motor oil
MO	moving out
mob	mobile

MOCT	moving out (off) cable tools
MOCU	moving out completion unit
mod	model
mod	moderate (ly)
mod	modification
modu	modular
MOE	milled other end
MOE	moving out equipment
Moen	Moenkopi
mol	molas
mol	mole
MOL	molecule, molecular
mol	mollusca
mol wt	molecular weight
mon	monitor
MON	motor octane number
Mont	Montoya
Moor	Mooringsport
MOP	maximum operating pressure
Mor	Morrow
MOR	moving out rig
Morr	Morrison
MORT	moving out (off) rotary tools
Mos	Mosby
mot	motor
mott	mottled
MOU	motor oil units
mov	moving
Mow	Mowry
MP	maximum pressure
MP	melting point
MP	multipurpose
mPa	megapascal
MPB	metal petal basket
MPGH-Lith	multipurpose grease lithium base
MPGR-Soap	multipurpose grease soap base
MPH	miles per hour

MPL	mechanical properties
MPT	male pipe thread
MPY	miles per year
MR	marine rig
MR	meter run
mr	milliroentgen
MRF	"merf"/main reaction furnace
MRK	marking
mrlst	marlstone
MRQ	memo requesting quotes
ms	millisecond (s)
MS	motor severe
MSA	multiple service acid
MSC	mapping subcommittee
Mscf	thousand standard cubic feet
Mscf/d	thousand standard cubic feet per day
Mscf/h	thousand standard cubic feet per hour
MSDS	material safety data sheets
msl	mean sea level
MSP	maximum surface pressure
MSR	mud/silt remover
mstr	master
MSW	muddy salt water
MSWG	miscible substance group
MT	empty container
MT	magnetic particle examination
MT	marcaroni tubing
Mt. Selm	Mount Selman
MTD	maximum total depth
MTD	mean temperature difference
MTD	measured total depth
mtd	mounted
mtg	mounting
mtge	mortgage

mtl	material
MTO	material take-off
MTP	maximum top pressure
MTP	maximum tubing pressure
mtr	meter
MTR	motor
MTS	mud to surface
Mtx	matrix
mud wt	mud weight
mudst	mudstone
MULTX	multiply, multiplexer
musc	muscovite
MUX	middle hydrocrackate
mv	millivolt
Mvde	Mesaverde
MVFT	motor vehicle fuel tax
MVOP	monthly volume operation plan
MW	megawatt
MW	microwave
MW	mud weight
MW	muddy water
MWD	marine wholesale distributors
MWP	maximum working pressure
MWPE	mill wrapped plain end
Mwy	Midway
mxd	mixed
M1E	milled one end
M2E	milled two ends
m/d	cubic meters per day

—————— **N** ——————

N	Newton
N	nonproducer
N	normal (to express concentration)

N	north
N. Cock.	Nonionella Cockfieldensis
N/O	north offset
N/S S/S	nonstandard service station
N/tst	no test
N/2	north half
N/4	north quarter
NA	not applicable
NA	not available
Nac	Nacotoch
nac	nacreous
NaCL	sodium chloride
NaCO₃	sodium carbonate
NAG	no appreciable gas
NALRD	Northern Alberta Land Registration District
NaOH	sodium hydroxide
nap	naphtha
NARR	narrative
nat	natural
nat'l	national
Nav	Navajo
Navr	Navarro
NB	new bit
NB	nitrogen blanket
Nbg	Newburg
NC	no change
NC	no core
NC	normally closed
NC	not completed
NCT	national coarse thread
NCT	noncontiguous tract
ND	nippled down
ND	nondetergent
ND	not drilling
NDBOPs	nipple (ed) (ing) down blowout preventers
NDE	not deep enough
NDG	no show gas
Ndl Cr	Noodle Creek

NDT	nipple-down tree
NDT	nondestructive testing
NE	nonemulsifying agent
NE	northeast
NE/4	northeast quarter
NEA	nonemulsion acid
NEC	National Electric Code
NEC	northeast corner
neg	negative
neg	negligible
NEGO	negotiation
NEL	northeast line
NEP	net effective pay
neut	neutral, neutralization
Neut. No.	Neutralization Number
New Alb	New Albany shale
Newc	Newcastle
NF	National Fine (thread)
NF	natural flow
NF	no fluid
NF	no fluorescence
NF	no fuel
NFD	new field discovery
NFOC	no fluorescence or cut
NFW	new field wildcat
NG	natural gas
NG	no gauge
NG	no good
NGL	natural gas liquids
NGTS	no gas to surface
NHDS	naphtha-hydrogen desulfurization
NH$_3$	ammonia
NH$_4$Cl	ammonium chloride
NIC	not in contract
NID	notice of intention to drill
Nig	Niagara
Nine	Ninnescah
Niob	Niobrara
nip	nipple

nitro	nitroglycerine
NL	north line
NL Gas	nonleaded gas
NLL	neutron lifetime log
N'ly	northerly
NMI	nautical mile
NO	new oil
NO	Noble-Olson
NO	normally open
NO	number
No Inc	no increase
no rec	no recovery
No.	number (before a number)
NOB	not on bottom
nod	nodule, nodular
Nod. blan.	Nodosaria blanpiedi
Nod. mex.	Nodosaria mexicana
NOJV	nonoperated joint ventures
nom	nominal
Non	Nonionella
nonf G	nonflammable compressed gas
NOP	nonoperating property
NOR	no order required
nor	normal
NOV	notice of violation
noz	nozzle
NP	nameplate
NP	nickel plated
NP	no production
NP	nonporous
NP	not prorated
NP	not pumping
NP	Notary Public
NPD	new pool discovery
NPDES	National pollution discharge elimination system
NPL	nipple
NPL UP	nippled up

npne	neoprene
NPOS	no paint on seams
NPR	Naval Petroleum Reserve
NPRA	Naval Petroleum Reserve, Alaska
NPS	nominal pipe size
NPSH	net positive suction head
NPT	National pipe thread
NPTF	National pipe thread, female
NPTM	National pipe thread, male
NPW	new pool wildcat
NPX	new pool exempt (nonprorated)
NR	new rod
NR	no recovery
NR	no report, not reported
NR	nonreturnable, no returns, not reached
NRC	Nuclear Regulatory Commission
NRI	net revenue interest
NRS	nonrising stem (valve)
NRSB	nonreturnable steel barrel
NRSD	nonreturnable steel drum
NS	no show
NSC	not suitable for coating
NSFOC	no show fluorescence or cut
NSO	no show oil
NSO&G	no show oil and gas
NSPS	new source performance standards
nstd	nonstandard
NT	net tons
NT	no time
NTD	new total depth
NTP	notice to proceed
NTS	not to scale
NU	naphfining unit
NU	nippled (ing) up

NU	nonupset
NUBOPs	nipple (ed) (ing) up blowout preventers
NUE	nonupset ends
Nug	nugget
num	numerous
NUT	nipple up tree
NUWH	nippling up wellhead
NVP	no visible porosity
NW	no water
NW	northwest
NW/C	northwest corner
NW/4	northwest quarter
NWL	northwest line
NWT	Northwest Territories
NYA	not yet available
NYD	not yet drilled
NYL	nylon
N$_2$	nitrogen

O

O	oil
O	Osborne
O sd	oil sand
O&G	oil and gas
O&GC SULW	oil and gas-cut sulfur water
O&GCAW	oil and gas-cut acid water
O&GCLW	oil and gas-cut load water
O&GCM	oil and gas-cut mud
O&GCSW	oil and gas-cut salt water
O&GCW	oil and gas-cut water
O&GL	oil and gas lease
O&M	operations and maintenance
O&SW	oil and salt water
O&SWCM	oil and sulfur water-cut mud

O&W	oil and water
O/S	out of service over and short (report)
O/S	out of stock
OA	overall
OAH	overall height
Oakv	Oakville
OAL	overall length
OAW	oil abandoned well
OB	off bottom
obj	object
OBM	oil-based mud
OBMO	outboard motor oil
OBS	observation
OBS	ocean bottom suspension
obsol	obsolete
OBW & RS	optimum bit weight and rotary speed
OC	oil cut
OC	on center
OC	open choke
OC	open cup
OC	operations commenced
OC-	Oil and Gas Investigations Chart
OCB	oil circuit breaker
occ	occasional (ly)
OCM	oil-cut mud
OCS	Outer Continental Shelf
OCSW	oil-cut salt water
oct	octagon, octagonal
oct	octane
OCW	oil-cut water
od	odor
OD	outside diameter
Odel	O'Dell
ODT	oil down to
OE	oil emulsion
OE	open end
OE	overexpenditure

OEB	other end beveled
OEM	oil emulsion mud
OF	open flow
OF	open-file report
off	office, official
off-sh	offshore
OFIC	oil insulated fan-cooled
OFL	overflush (ed)
OFLU	oil fluorescence
OFOE	orifice flange one end
OFP	open flow potential
OFS	offsite
OGJ	Oil & Gas Journal
O'H	O'Hara
OH	open hearth
OH	open hole
OH	overhead
ohm	ohm
ohm-cm	ohm-centimeter
ohm-m	ohmmeter
OI	oil insulated
OIFC	oil insulated, fan cooled
OIH	oil in hole
Oil Cr	Oil Creek
oilfract	oil fractured
OIP	oil in place
OISC	oil insulated, self-cooled
OIT	oil in tanks
OIWC	oil immersed, water cooled
OL	off/on location
OL	open line (no choke)
ole	olefin
Olig	Oligocene
OLN	outline
OMRL	oriented microresistivity
ONR	octane number requirement
ONRI	octane number requirement increase
OO	oil odor
ooc	ooliclastic

OOIP	original oil in place
ool	oolitic
oom	oolimoldic
OP	articles published in outside journals/books
OP	oil pay
OP	outpost
OP	overproduced
op hole	open hole
OPB	old plugback
OPBD	old plugback depth
oper	operate, operations, operator
Operc	Operculinoides
OPI	oil payment interest
opn	open (ed) (ing)
OPO	overseas procurement office
opp	opposite
OPT	official potential test
OPTL	optional
optn to F/O	option to farmout
OR	orange
Or	Oread
Ord	Ordovician
orf	orifice
org	organic
org	organization
ORIAC	Oil Refining Industry Action Committee
ORIENT	orientation
orig	original, originally
Orisk	Oriskany
ORR	overriding royalty
ORRI	overriding royalty interest
ORSANCO	Ohio River Valley Water Sanitation Commission
orth	orthoclase
OS	oil show
Os	Osage
OS	overshot

OS&F	odor stain and fluorescence
OS&Y	outside screw and yoke (valve)
OSA	oil-soluble acid
OSD	operation shutdown
OSF	oil string flange
OSI	oil well shut in
OSIDP	oil standing in drillpipe
Ost	Ostracod
OSTN	oil stain
OSTOIP	original stock tank oil in place
Osw	Oswego
OT	open tubing
OT	overtime
OT&S	odor taste & stain
OTD	old total depth
OTD	original total depth
OTE	oil-powered total energy
otl	outlet
OTS	oil to surface
OTS&F	odor taste stain and fluorescence
OU	oil unit
Our	Ouray
ovhd	overhead
OWC	oil-water contact
OWDD	old well drilled deeper
OWF	oil well flowing
OWFWF	oil well from waterflood
OWG	oil well gas
OWPD	old well plugged back
OWST	old well sidetracked
OWWO	old well worked over
ox	oxidized, oxidation
oxy	oxygen
oz	ounce

P

P	professional paper
P & A	plugged and abandonded
P & ID	process & instrument diagram
P & NG	petroleum and natural gas
P Lar	Post Laramie
P tstg	pump testing
p.	page (before a number)
P.O.	Pin Oak
P.O.	Post Oak
P.P.	present production
P-HDII	perforating Hyperdome II
P-M	Pensky-Martins (flash)
P&C	personal and confidential
P&F	pump and flow
P&IDS	piping and instrument diagrams
P&L	profit and loss
P&P	porosity and permeability
P&P	porous and permeable
P/	pump
P/BLDG	pump building
P/DIA	piping diagram
PA	participating area
Pa	Pascal
PA	pooling agreement
PA	pressure alarm
PA	public address
PAB	per-acre bonus
Padd	Paddock
Paha	Pahasapa
Pal	Paluxy
Paleo	paleontology
Paleo	Paleozoic
Palo P	Palo Pinto
Pan L	Panhandle lime
PAR	per-acre rental
Para	Paradox

Park C	Park City
PART	partial
pat	patent (ed)
patn	pattern
pav	paving
Paw	Pawhuska
PB	plugged back
PB-ADA	report available only through National Technical Information Service
PBD	plugged-back depth
PBE	plain both ends
PBHL	proposed bottom-hole location
pbl (y)	pebble, pebbly
PBP	pulled bid pipe
PBTD	plugged back total depth
PBW	pipe buttweld
PBX	private branch exchange
PBX	switchboard
PC	Paint Creek
pc	piece
PC	poker chipped
pc	port collar
PC	Porter Creek
PCF	pounds per cubic foot
pct	percent
PCV	positive crankcase ventilation
PCV	pressure control valve
PD	geopressure development, failure
pd	paid
PD	per day
PD	plug down
PD	present depth
PD	pressed distillate
PD	proposed depth
PD	pulsation dampener

PD	pumper's depth
PDC	power distribution center
PDC	pressure differential controller
PDET	production department exploratory test
PDI	pressure differential indicator
PDIC	pressure differential indicator controller
PDR	pressure differential recorder
PDRC	pressure differential recorder controller
PDS	geopressure development, success
PDS	power distribution system
pdso	pseudo
pe	pin end
PE	plain end
PE	pumping equipment
PEB	plain end beveled
PED	pedestal
pell	pelletal, pelletoidal
pen	penetration, penetration test
Pen A.C.	penetration asphalt cement
penal	penalty, penalize (ed) (ing)
Penn	Pennsylvanian
perco	percolation
perf	perforate (ed) (ing) (or)
perf csg	perforated casing
perm	permanent
perm	permeable (ability)
Perm	Permian
perp	perpendicular
pers	personnel
PERT	performance evaluation and review technique
pet	petroleum

Pet	Pettet
Pet sd	Pettus sand
petrf	petroliferous
petrochem	petrochemical
Pett	Pettit
PEW	pipe electric weld
pf	per foot
PF	power factor
pfd	preferred
PFD	process flow diagram
PFM	power factor meter
PFRACT	prefractionator
PFT	pumping for test
PG	Pecan Gap
Pg	plug
PGC	Pecan Gap chalk
PGW	producing gas well
pH	acidity or alkalinity
pH	hydrogen ion concentration
pH	measure of hydrogen potential
Ph	parish
ph	phase
PHC	pipe-handling capacity
Phos	Phosphoria
PI	penetration index
PI	Pine Island
PI	pressure indicator
PI	productivity index
PI	pump in
PIC	pressure indicator controller
Pic Cl	Pictured Cliff
pinpt	pinpoint
PIP	pump-in pressure
piso	pisolites, pisolitic
pit	pitted
PJ	pump jack
PJ	pump job
pk	pink

Pk Lkt	Point Lookout
pkg (d)	packing, package (ed)
pkr	packer
PL	pipeline
PL	plate
PL	property line
pl fos	plant fossils
Plan. palm.	Planulina palmarie
Plan. har.	Planulina harangensis
plas	plastic
PLASR	plaster
platf	olatform
plcy	pelecypod
pld	pulled
PLE	plain large end
Pleist	Pleistocene
plg	plagioclase
plg	pulling
plgd	plugged
Plio	Pliocene
PLMB	plumbing
pln	plan
plngr	plunger
PLO	pipeline oil
PLO	pumping load oil
plt	pilot
PLT	pipeline terminal
plt	plant
plty	platy
PLV	pilot loaded valve
PLW	pipe lapweld
P-M	Pensky Martins
PML	production management
pmp (d) (g)	pump (ed) (ing)
PN	Performance Number (aviation gas)
pneu	pneumatic
pnl	panel
PNL BD	panel board
PNR	please note and return

po	Phrohotite
PO	pulled out
PO	pumps off
PO	purchase order
POB	plug on bottom
POB	pump on beam
POCS	Pacific Outer Continental Shelf
POD	plan of development
Pod.	Podbielniak
POE	plain one end
POGW	producing oil and gas well
POH	pulled (put) out of hole
pois	poison
pol	polish (ed)
poly	polymerization, polymerized
poly cl	polyethylene
polygas	polymerized gasoline
polypl	polypropylene
PONA	parrafins-olefins-napthenes-aromatics
Pont	Pontotoc
POOH	pull (put) out of hole
POP	putting on pump
por	porosity, porous
porc	porcelaneous
porc	porcion
port	portable
pos	position
pos	positive
poss	possible (ly)
pot	potential
pot dif	potential difference
POT/WTR	potable water
pour ASTM	pour point (ASTM method)
POW	producing oil well
POWF	producing oil well, flowing
POWP	producing oil well, pumping
PP	pinpoint

PP	production payment
PP	pulled pipe
PP	pump pressure
ppb	parts per billion
PPB	pounds per barrel
ppd	prepaid
ppg	piping
PPG	pounds per gallon
PPI	production payment interest
PPI	Process Performance Index
ppm	parts per million
ppn no	precipitation number
PPP	pinpoint porosity
ppt	precipitate
pr	pair
PR	polished rod
pr	poor
PR	pressure recorder
PR	public relations
PR	purchasing request
pr op	present operations
PR&T	pull (ed) rods and tubing
PRC	pressure recorder control
prcst	precast
prd	period
Pre Camb	Precambrian
PRECIP	precipitator
predom	predominant
prefab	prefabricated
prehtr	preheater
prelim	preliminary
prem	premium
Prep	prepare, preparing, preparation
press	pressure
prest	prestressed
prev	prevent, preventive
PREV	previous
PREV DO AVG	previous daily average

PRF	Primary Reference Fuel
pri	primary
prin	principal
pris	prism (atic)
priv	privilege
prly	pearly
prmt	permit
prncpl lss	principal lessee (s)
pro	prorated
prob	probable (ly)
proc	process
prod	produce (ed) (ing) (tion), product (s)
prog	progress
proj	project (ed) (ion)
PROP	property
prop	proportional
prop	propose (ed)
prot	protection
Protero	Proterozoic
Prov	provincial
PRPT	preparing to take potential test
prtgs	partings
PS	pressure switch
ps	pseudo
PS	pump station
PSA	packer set at
PSB	pressure seal bonnet
PSD	permanently shut down
PSD	prevention of significant deterioration, EPA
PSE	plain small end
psf	pounds per square foot
psi, PSI	pounds per square inch
PSI	profit-sharing interest
psia, PSIA	pounds per square inch absolute
psig, PSIG	pounds per square inch gauge
PSL	pipe sleeve

PSL	Public School Land
PSM	pipe seamless
Psp	prospect
PSV	pressure safety valve
PSW	pipe spiral weld
PT	liquid penetrant examination
pt	part, partly
pt	pint
pt	point
PT	potential test
PTC	permanent type completion
PTD	present total depth
PTD	projected total depth
PTD	proposed total depth
PTF	production test flowed
PTG	pulling tubing
PTN	partition
PTP	production test pumped
PTR	pulling tubing and rods
PTS pot	pipe to soil potential
PTTF	potential test to follow
PU	picked up
PU	pulled up
PU	pumping unit
PUC	project ultimate cost
PUDP	picking up drillpipe
PUIC	pulled up in casing
PULS	pulse (sating) (sation)
PURCH	purchasing
PURF	purification
purp	purple
PV	plastic viscosity
PV	pore volume
PVC	polyvinyl chloride
pvmnt	pavement
PVR	plant (pressure) volume reduction
PVT	pressure-volume-temperature

PW	producing well
PW(15)	present worth at discount rate of 15%
PWHT	postweld heat treatment
PWR	power
PWY	pipeway
PWZ	peripheral wedge zone
Pxy	Paluxy
pyls	pyrolysis
pymt	payment
pyr	pyrite, pyritic
pyrbit	pyrobitumen
pyrclas	pyroclastic

—— Q ——

Q. City	Queen City
Q. sd	Queen Sand
QA	quality assurance
QC	quality control
QDA	quality discount allowance
QDRNT	quadrant
qnch	quench
QRC	quick ram change
qry	quarry
qt	quart (s)
qtr	quarter
qty	quantity
qtz	quartz, quartzite, quartzitic
qtzose	quartzose
quad	quadrant, quadrangle, quadruple
QUAL	qualitative
qual	quality
quan	quantity
quest	questionable
quint	quintuplicate

R

R	radius
R	range
R	rankine (temp. scale)
R	resistivity
r	roentgen
R	rows
R & D	research and development
R & T	rods and tubing
R test	rotary test
R.O.	Red Oak
R(16")	resistivity as recorded from 16" electrode configuration
R-SP	recommended spare part
R&L	road & location
R&LC	road & location complete
R&O	rust and oxidation
R/A	regular acid
R/W	right of way
RA	radioactive
RA	right angle
RACTR	reactor
rad	radical
rad	radiological
rad	radius
RADT	radiant
radtn	radiation
RAGL	raw gas lift
RALOG	running radioactive log
Rang	Ranger
RB	rock bit
RB	rotary bushing
RBLR	reboiler
Rbls	rubber balls
RBM	rotary bushing measurement
RBP	retrievable bridge plug
rbr	rubber

RBSO	rainbow show of oil
RBSOF	rubber ball sand oil frac
RBSWF	rubber ball sand water frac
RC	rapid curing
RC	Red Cave
RC	remote control
RC	reverse circulation
RC	running casing
RCO	returning circulation oil
RCPT	receptacle
RCR	Ramsbottom Carbon Residue
RCR	reverse circulation rig
RCTN	reaction
RCVR	receiver
RCVY	recovery
RCYL	recycle
RD	redrilled
RD	rigged (ing) down
rd	road
rd	round
Rd Bds	red beds
Rd Fk	Red Fork
Rd Pk	Red Peak
rd thd	round thread
RDB	rotary drive bushing
RDB-GD	rotary drive bushing to ground
RDCR	reducer
rdd	rounded
RDMO	rigged down, moved out
RDS	reservoir description service
RDSU	rigged-down swabbing unit
rdtp	round trip
REABS	reabsorber
reacd	reacidize (ed) (ing)
React	reaction (ed)
rebar	reinforcing bar
rec	recommend
rec	record (er) (ing)

rec	recover (ed) (ing), recovery
recd	received
recip	reciprocate (ing)
recirc	recirculate
recomp	recomplete (ed) (ion)
RECOMP	recompressor
recond	recondition (ed)
recp	receptacle
rect	rectangle, rectangular
rect	rectifier
recy	recycle
red	reducing, reducer
red bal	reducing balance
redrld	redrilled
ref	reference
ref	refine (ed) (er) (ry)
refg	refining
refl	reflection
refl	reflux
REFMR	reformer
REFOO	re-evaluation for overoptimism
reform	reformate (er) (ing)
refr	refraction, refractory
REFRIG	refrigerator (rant) (tion)
reg	register
reg	regular, regulator
regen	regenerator
reinf	reinforce (ed) (ing) (ment)
reinf conc	reinforced concrete
rej	reject
rej'n	rejection
Rek	Reklaw
rek	rock
rel	relay
rel	release (ed)
rel	rig released
REL	running electric log
reloc	relocate (ed)
rem	remains

rem	remedial
Ren	Renault
rent	rental
Reo. bath.	Reophax bathysiphoni
rep	repair (ed) (ing) (s)
rep	replace (ed)
rep	report
reperf	reperforated
repl	replace (ment)
REQ	request
req	requisition
reqd	required
reqmt	requirement
res	research
res	reserve (ation)
res	resistance, resistivity, resistor
Res. O. N.	Research Octane Number
resid	residual, residue
RESIS	resistor (s)
ret	retain (er) (ed) (ing)
ret	return
retd	returned
retr	retrievable
retr ret	retrievable retainer
rev	reverse (ed)
rev	revise (ed) (ing) (ion)
rev	revolution (s)
rev/O	reversed out
RF	raised face
RF	rig floor
RFFE	raised face, flanged end
RFG	roofing
RFG/BD	refrigeration building
RFLCT	reflect (ed) (ing) (tion)
RFP	request for proposal
RFQ	request for quote
RFR	ready for rig
RFSF	raised face, smooth finish
RFSO	raised face, slip on

RFWN	raised face, weld neck
RG	raw gas
rg	ring
RG	ring groove
Rge	range
rgh	rough
RGTR	register
RH	rat hole
RH	relative humidity
RH	right hand
RHD	righthand door
rheo	rheostat
RHM	rat hole mud
RHN	Rockwell hardness number
RI	royalty interest
Rib	ribbon sand
Rier	Rierdon rig
RIH	ran in hole
RIL	red indicating lamp
riv	rivet
RIZ	resistivity invaded zone
RJ	ring joint
RJFE	ring joint, flanged end
RK	rack
rk	rock
RKB	rotary kelly bushing
rky	rocky
RL	random lengths
rlf	relief
rlg	railing
rls (d) (ing)	release (ed) (ing)
rly	relay
rm	ream
Rm	resistivity, mud
rm	room
rmd	reamed
Rmf	resistivity, mud filtrate
rmg	reaming
rmn	remains
RMS	root mean square

rmv (l)	remove (al) (able)
rnd	rounded
rng	running
RO	reversed out
ro	rose
Ro	Rosiclare sand
Rob	Robulus
Rod	Rodessa
ROF	rich oil fractionator
ROGL	rotative gas lift
ROI	return on investment
ROL	rig on location
ROM	rough order of magnitude
ROM	run of mine
RON	Research Octane Number
ROP	rate of penetration
ROR	rate of return
ROS	remote operating system (station)
rot	rotary, rotate, rotator
ROW	right of way
roy	royalty
RP	rock pressure
RPI	Research Planning Institute
rpm	revolutions per minute
rpmn	repairman
RPP	retail pump price
RPRT	report
rps	revolutions per second
rptd	reported
RR	railroad
RR	Red River
RR	rig released
RR	rig repair
RR	rigging rotary
RR&T	ran (running) rods and tubing
RRC	Railroad Commission (Texas)
RS	rig service

RS	rig skidded
RS	rising stem (valve)
RSD	returnable steel drum
rsns	resinous
RSU	released swab unit
rsvr	reservoir
RT	radiographic examination
RT	rig time
RT	rotary table
RT	rotary tools
RT CB	round trip changed bit
rtd	retard (ed)
RTD	rotary total depth
rtg	rating
RTG	routing
RTG	running tubing
RTJ	ring tool joint
RTJ	ring-type joint
RTL	Refinery Technology Laboratory
RTLTM	rate too low to measure
RTMTR	rotameter
rtnr	retainer
RTTS	retrievable test treat squeeze (tool)
RTU	remote terminal unit
RU	rig (ged) (ging) up
RU	rotary unit
rub	rubber
RUCC	rig-up casing crew
RUCT	rigging-up cable tools
RUM	rigging-up machine
RUP	rigging-up pump
rupt	rupture
RURT	rigging-up rotary tools
RUSR	rigging-up service rig
RUST	rigging-up standard tools
RUSU	rigging-up swabbing unit
RUT	rigging-up tools
RV	relief valve

RVP	Reid vapor pressure
rvs (d)	reverse (ed)
RVT	rivet
Rw	resistivity, water
Rwa	resistivity, water (apparent)
rwk (d)	rework (ed)
RWTP	returned well to production
Rxo	resistivity, flushed zone

S

S	seconds
S	south
S	stratigraphic test
s & s	spigot and spigot
S & T	shell and tube
S Bomb	sulfur by bomb method
S O	south offset
S Riv	Seven Rivers
S.L.	sea level
S-T-R	section-township-range
S&F	swab and flow
S&O	stain and odor
s&p	salt and pepper
S/	swabbed
S/C	speed/current
S/E	screwed end
S/FAB	shop fabrication
s/p	shipping point (purchasing term)
S/SR	sliding-scale royalty
S/SW	screwed and socketweld
S/T	sample tops
S/T	speed/torque
S/WTR	sanitary water
S/2	south half
s,t&b	sides, tops, & bottoms
SA	seal assembly

Sab	Sabinetown
sach	saccharoidal
Sad Cr	Saddle Creek
sadl	saddle
saf	safety
SAF/DPT	safety/department
SAFE	surface approximation and formation evaluation
Sal	Salado
sal	salary, salaried
sal	salinity
Sal Bay	Saline Bayou
salv	salvage
samp	sample
SAN	sanitary
San And	San Andres
San Ang	San Angelo
San Raf	San Rafael
Sana	Sanastee
sani	sanitary
sap	saponification
Sap No.	saponification number
Sara	Saratoga
sat	saturated, saturation
Saw	Sawatch
Sawth	Sawtooth
Say Furol	Saybolt furol
SB	sideboom
SB	sleeve bearing
SB	stuffing box
sb	sub
Sb	Sunburst
SBA	secondary butyl alcohol
SBB&M	San Bernardino base and meridian
SBHP	static bottom-hole pressure
sc	scales
SC	self-contained
SC	show condensate
SC DL	slip and cut drill line

SCAF	scaffolding
scatt (d)	scatter (ed)
scf, SCF	standard cubic foot
scfd, SCFD	standard cubic feet per day
scfh, SCFH	standard cubic feet per hour
scfm, SCFM	standard cubic feet per minute
sch	schedule
schem	schematic
SCHL	Schlumberger
scly	securaloy
scolc	scolescodonts
scr	scraper
scr	scratcher
scr	screen
scr (d)	screw (ed)
scrub	scrubber
SCSSV	surface-controlled subsurface safety valve
sctrd	scattered
sd, SD	sand
SD	shut down
sd & sh	sand and shale
SD Ck	side door choke
Sd SG	sand showing gas
Sd SO	sand showing oil
SDA	shut down to acidize
sdd	sanded
SDF	shut down to fracture
sdfract	sandfracked
SDG	siding
SDL	shut down to log
SDO	show of dead oil
SDO	shut down for orders
sdoilfract	sand-oil fracked
SDON	shut down overnight
SDP	set drillpipe
SDPA	shut down to plug and abandon
SDPL	shut down for pipeline

SDR	shut down for repairs
sdtkr	sidetrack (ed) (ing)
SDW	shut down for weather
SDWL	sidewall
SDWO	shut down awaiting orders
sdwtrfract	sand-water fracked
sdy	sandy
sdy li	sandy lime
sdy sh	sandy shale
SE	southeast
SE NA	screw end American National Acme thread
SE NC	screw end American National Coarse thread
SE No.	steam emulsion number
SE NTP	screw end American National Taper Pipe thread
SE/C	southeast corner
SE/4	southeast quarter
Sea	Seabreeze
sec	secant
sec	second (ary)
sec	secretary
sec	section
SECT	section (s) (al) (ing)
sed	sediment (s)
Sedw	Sedwick
SEG	segment
seis	seismograph, seismic
sel	selenite
Sel	Selma
SELECT	selection (tive) (tor)
Sen	Senora
SEO	seal oil
SEP	separate, separator, separation
SEPM	Society of Economic Paleontologists & Mineralogists

sept	septuplicate
seq	sequence
ser	series, serial
Serp	serpentine
Serr	Serratt
serv	service (s)
serv chg	service charge
set	settling
sew	sewer
SEWOP	self-elevating work platform
Sex	sexton
sext	sextuplicate, sextuplet
SF	sandfrac
sfc	surface
SFD	system flow diagram
SFL	starting fluid level
SFLU	slight, weak, or poor fluorescence
SFO	show of free oil
SFP	surface flow pressure
SFT	sample formation tester
sft	soft
sg	specific gravity
SG	show gas
SG	surface geology
SG & W	show gas and water
SG&C	show gas and condensate
SG&D	show gas and distillate
SG&O	show of gas and oil
SGCM	slightly gas-cut mud
SGCO	slightly gas-cut oil
SGCSW	slightly gas-cut salt water
SGCW	slightly gas-cut water
SGCWB	slightly gas-cut water blanket
SGCWC	slightly gas-cut water cushion
sgd	signed
sgl (s)	single (s)
sh	shale

sh	sheet
SH	substructure height
Shan	Shannon
SHDP	slim-hole drillpipe
Shin	Shinarump
shld	shoulder
shls	shells
SHLT	shelter
shly	shaly
SHM	solar heat medium
shp	shaft horsepower
shp(g)	ship (ping)
shpt	shipment
shr	shear
SHT	straight-hole test
SHTG	sheeting
SHTG	shortage
shthg	sheathing
SI	shut in
SIBHP	shutin bottom-hole pressure
SICP	shutin casing pressure
SIGW	shutin gas well
SIH	started in hole
Sil	Silurian
silic	silica, siliceous
silt	siltstone
sim	similar
Simp	Simpson
SIOW	shutin oil well
SIP	shutin pressure
Siph. d.	Siphonina davisi
SITP	shutin tubing pressure
SIWHP	shutin wellhead pressure
SIWOP	shutin, waiting on potential
sk	sack (s)
SK	sketch
Sk Crk	Skull Creek
skim	skimmer
Skn	Skinner
skr rd	sucker rod

sks	slickensided
skt	socket
SL	section line
sl	sleeve
SL	south line
SL	state lease
SLC	steel line correction
sld	sealed
sli	slight (ly)
Sli	Sligo
sli SO	slight show of oil
slky	silky
SLM	steel line measurement
SLNCR	silencer
slnd	solenoid
SLPR	sleeper
Slt Mtn	Salt Mountain
Slty	salty
slty	silty
slur	slurry
SLV	sleeve
S'ly	southerly
SM	Seward Meridian (Alaska)
sm	small
SM	surface measurement
Smithw	Smithwick
Smk	Smackover
smls	seamless
smpl	sample
smth	smooth
SN	seating nipple
SNG	synthetic natural gas
SNUB	snubber, snubbing
SNUFF	snuffing
SO	shake out
SO	shaled out
SO	show of oil
SO	side opening
SO	slip on
SO&G	show oil and gas

SO&GCM	slightly oil- and gas-cut mud
SO&W	show oil and water
SOC	start of cycle
SOCM	slight oil-cut mud
SOCSW	slightly oil-cut salt water
SOCW	slight oil-cut water
SOCWB	slightly oil-cut water blanket
SOCWC	slightly oil-cut water cushion
sod gr	sodium-base grease
SOE	screwed on one end
SOF	sand-oil fracture
SOH	shot open hole
SOH	started out of hole
sol	solenoid
sol	solids
soln	solution
solv	solvent
som	somastic
somct	somastic coated
SONL	sonic log
SOOH	started out of hole
SOOH	strapped out of hole
SOP	standard operational procedure
SOR	start of run
sort	sorted (ing)
SOV	solenoid-operated valves
sow	socket weld
SP	self (spontaneous) potential
SP	set plug
sp	shipping point (purchasing term)
sp	shot point
sp	slightly porous
sp	spare
Sp	Sparta
sp	spore

SP	straddle packer
SP	surface pressure
sp	shipping point (purchasing term)
sp gr	specific gravity
sp ht	specific heat
SP SW	single pole switch
sp. vol.	specific volume
SP-DST	straddle packer drillstem test
spcl	special
spcr	spacer
spd	spud (ded) (der)
spdl	spindle
SPDT	single-pole double throw
SPDT SW	single-pole double throw switch
SPE	Society of Petroleum Engineers
spec	specification
speck	speckled
SPF	shot per foot
spf	spearfish
spg	sponge
spg	spring
SPGT	spigot
sph	spherules
Sphaer	Sphaerodina
sphal	sphalerite
spic	spicule (ar)
Spiro. b.	Spiroplectammina barrowi
spkr	sprinkler
spkt	sprocket
spl	sample
spl cham	sample chamber
Spletp	Spindletop
SPLTR	splitter
splty	specialty
splty	splintery
sply	supply

SPM	strokes per minute
Spra	Spraberry
sprf	spirifers
Sprin	Springer
SPST	single-pole single throw
SPST SW	single-pole single throw stitch
SPT	shallower pool (pay) test
sptd	spotted
sptty	spotty
Spud Date	date actually started drilling
SPWY	spillway
sq	square
sq cg	squirrel cage
sq cm	square centimeter
sq ft	square foot
sq in.	square inch
sq km	square kilometer
sq m	square meter
sq mm	square millimeter
sq pkr	squeeze packer
sq yd	square yard (s)
SQRT	square root
sqz	squeeze (ed) (ing)
SR	short radius
SR	swab rate
SR	swab run (s)
SRB	sulfate bacteria
SRG	surge
SRL	single random lengths
SRN	straight-run naphtha
srt	sort (ed) (ing)
SRV	safety relief valve
SS	sandstone
SS	service station
SS	shock sub
SS	short string
SS	single shot
SS	slow set (cement)

SS	small show
SS	stainless steel
SS	string shot
SS	subsea
SS	subsurface
SSA	spot sales agreement
SSG	slight show of gas
SSO	slight show of oil
SSO&G	slight show of oil and gas
SSSV	subsurface safety valve
SSU	Saybolt Seconds Universal
SSUW	salty sulfur water
ST	short thread
st	start
ST (g)	sidetrack (ing)
St L	Saint Louis lime
St Ptr	Saint Peter
St Gen	Saint Genevieve
ST&C	short threads & coupling
sta	station
Sta Marg	Santa Margarita
stab	stabilized (er)
STAG	staggered
Stal	Stalnaker
Stan	Stanley
stat	stationary
stat	statistical
State pot	state potential
stb, STB	stock tank barrels
stb/d, STBPD	stock tank barrels per day
stcky	sticky
std	stand (s) (ing)
std (s) (g)	standards
stdg	standing
stdy	steady
stdy	study
STDZN	standardization
Stel	Steele
stenp	stenographer
Stens	Stensvad

STG	stage
stging	straightening
STH	sidetracked hole
STIF	stiffener
stip	stippled
stir	stirrup
stk	stock
stk	streak (s) (ed)
stk	stuck
stl	steel
stm	steam
STM	steel tape measurement
stm cyl oil	steam cylinder oil
stm eng oil	steam engine oil
STM TR	steam trace (ing)
stn (d) (g)	stain (ed) (ing)
Stn Crl	Stone Corral
stn/by	stand by
stncl (d) (g)	stencil (ed) (ing)
Stnka	Satanka
stnr	strainer
stoip	stock tank oil in place
stor	storage
STP	standard temperature and pressure
STPR	strip (per) (ping)
stpr (d)	stopper (rd)
Str	Strawn
strat	stratigraphic
strd	straddle
strd	strand (ed)
strg	storage
strg	strong
strg (r)	string (er)
stri	striated
strk	streak
STROH	strap out of hole
strom	stromatoporoid
strt	straight
strtd	straightened

struc	structure, structural
STTD	sidetracked total depth
STV	stock tank vapor
stv	stove oil
stwy	stairway
Sty Mtn	Stony Mountain
styo	styolite, styolitic
sub	subsidiary
sub	substance
sub angl	subangular
Sub Clarks	sub-Clarksville
sub rnd	subrounded
subd	subdivision
SUBST	substitute
substa	substation
suc	sucrose, sucrosic
suct	suction
sug	sugary
sul	sulfur
sul wtr	sulfur water
SULF	sulfur, sulfuric
sulf	sulfated
SUM	summarize
sum	summary
Sum	Summerville
Sunb	Sunburst
Sund	Sundance
Sup	Supai
supl	supply (ied) (ier) (ing)
supp	supplement
suppt	support
suprv	supervisor
supsd	superseded
supt	superintendent
sur	survey
surf	surface
surp	surplus
SUS	Saybolt universal seconds
susp	suspended
SUSP CLG	suspended ceiling

SUV	Saybolt universal viscosity
SV	solenoid valve
svc	service
svcu	service unit
SVI	smoke volatility index
Svry	severy
SW	salt wash
SW	salt water
SW	socket weld
SW	southwest
SW	spiral weld
SW	switch
SW&W	show gas and water
SW/c	southwest corner
SW/4	southwest quarter
Swas	Swastika
SWB	seal-welded bonnet
SWB	swabbed, swabbing
swbd	switchboard
SWC	sidewall cores
SWCM	saltwater-cut mud
SWD	saltwater disposal
swd	swaged
SWDS	saltwater disposal system
SWDW	saltwater disposal well
Swet	sweetening
SWF	saltwater fracture
swgr	switchgear
SWI	saltwater injection
SWION	shut well in overnight
SWLD	seal weld
SWNP	sidewall neutron porosity
SWP	steam working pressure
SWR	statewide rules
SWRK	switchrack
SWS	sidewall samples
swtr	salt water
SWTS	salt water to surface
SWU	swabbing unit
sx	sacks

sxtu	sextuple
Syc	Sycamore
Syl	Sylvan
sym	symbol
sym	symmetrical
syn	synchronous, synchronizing
syn	synthetic
syn conv	synchronous converter
SYNSCP	synchroscope
SYNTH	synthesis
sys	system
sz	size

T

T	tee
T	tooth, teeth
T	ton (after number–3T)
T	township (as T2N)
T & B	top and bottom
T & C	threaded and coupled
T & C	topping and coking
T.O.	temperature observation
T&BC	top and bottom chokes
T&G	tongue and groove (joint)
T&R	tubing and rods
T&W	tarred and wrapped
T/	top of (a formation)
T/Box	terminal box
T/BRD	terminal board
T/C	tank car
T/C	turbine compressor
T/pay	top of pay
T/S	top salt
T/sd	top of sand
TA	temporarily abandoned
TA	turn around
tab	tabular, tabulating
TACH	tachometer

Tag	Tagliabue
Tal	Tallahatta
Tamp	Tampico
TAN	tangent
Tan	Tansill
Tann	Tannehill
TAPS	Trans-Alaska Pipeline System
Tark	Tarkio
Tay	Taylor
TB	tank battery
TB	thin bedded
tb	tube
TB/BDL	tube bundle
TBA	tertiary butyl alcohol
TBA	tires, batteries, and accessories
TBE	threaded both ends
tbg	tubing
tbg chk	tubing choke
tbg press	tubing pressure
TBP	true boiling point
TC	temperature controller
TC	tool closed
TC	top choke
TC	tubing choke
TCC	tag closed cup (flash)
TCC	thermofor catalytic cracking
TCC	tubing and casing cutter
Tcf, TCF	trillion cubic feet
Tcf/d, TCF/D	trillion cubic feet per day
TCP	tricresyl phosphate
TCV	temperature control valve
TD	time delay
TD	total depth
TDA	temporary dealer allowance
TDI	temperature differential indicator
TDR	temperature differential recorder

TDT	thermal decay time
TE	temporary
tech	technical, technician
TEFC	totally enclosed, fan cooled
tel	telephone, telegraph
TEL	tetraethyl lead
Tel Cr	Telegraph Creek
Temp	temperature
temp	temporary (ily)
Tens	Tensleep
tens str	tensile strength
Tent	tentaculites
tent	tentative
Ter	tertiary
termin	terminate (ed) (ing) (ion)
Tex	Texana
tex	texture
Text. art.	Textularia articulate
Text. d.	Textularia dibollensis
Text. h.	Textularia hockleyensis
Text. w.	Textularia warreni
TFB	trip (ped) for bit
Tfing	Three Finger
Tfks	Three Forks
TFNB	trip for new bit
tfs	tuffaceous
TG	temperature gradient
th	thence
TH	tight hole
Thay	Thaynes
THC	top hole choke
THD	thermal hydrodealkylation
thd	thread, threaded
Ther	Thermopolis
therm	thermometer
therm ckr	thermal cracker
therst	thermostat
THF	tubinghead flange
THFP	top hole flow pressure
thk	thick, thickness

thrling	throttling
thrm	thermal
thru	through
Thur	Thurman
TI	temperature indicator
ti	tight
TIC	temperature indicator controller
TIH	trip in hole
TIM	Timpas
Timpo	Timpoweap
tk	tank
TKF	tank farm
tkg	tankage
tkr	tanker (s)
tl	tool (s)
tl jt	tool joint
TLE	thread large end
TLG	telegraph
TLH	top of liner hanger
TML	tetramethyl lead
tndr	tender
TNS	tight no show
TO	temperature observation
TO	tool open
TOBE	thread on both ends
TOC	tag open cup (flash)
TOC	top of cement
TOCP	top of cement plug
Tod	Todilto
TOE	threaded one end
TOF	top of fish
TOH	trip out of hole
tol	tolerance
TOL	top of liner
tolu	toluene
Tonk	Tonkawa
tons	tons
TOP	testing on pump
topg	topping

topo	topographic, topography
TOPS	turned over to producing section
Tor	Toronto
Toro	Toroweap
TORT	tearing out rotary tools
TOS	top of salt
tot	total
Tow	Towanda
TP	toolpusher
TP	Travis Peak
TP	treating pressure
TP	tubing pressure
TP&A	theoretical production and allocation
TPC	top of cement
TPC	tubing pressure, closed
TPF	threaded pipe flange
TPF	tubing pressure, flowing
tpk	turnpike
Tpka	Topeka
TPSI	tubing pressure shut in
TR	temperature recorder
TR	trace
TR	tract
trans	transfer (ed) (ing)
trans	transformer
trans	transmission
transl	translucent
transp	transparent
transp	transportation
TRC	temperature recorder controller
Tremp	Teremplealeau
Tren	Trenton
TRG	to be conditioned for gas
Tri	Triassic
trilo	trilobite
Trin	Trinidad
trip	triplicate

Trip	Tripoli
trip	tripolitic
trip	tripped (ing)
trk	truck
trkg	trackage
Trn	Trenton
TRNDC	transducer
TRO	to be conditioned for oil
TRQ	torque
trt (r)(d)(g)	treat (er) (ed) (ing)
trtr	treater
TRVL	travel (ed) (ing)
TS	tensile strength
TS	topo sheet evaluation
TSD	temporarily shut down
TSE	thread small end
TSE-WLE	thread small end, weld large end
TSI	temporarily shut in
TSITC	temperature survey indicated top cement at
TSS	Tar Springs sand
tst (r) (g)	test (er) (ing)
tste	taste
TSTM	too small to measure
TT	tank truck
TT	through-tubing
TTC	through-tubing caliper
TTF	test to follow
TTL	total time lost
TTP	through-tubing plug
TTTT	turned to test tank
Tuck	Tucker
tuf	tuffaceous
Tul Cr	Tulip Creek
tung carb	tungsten carbide
TURB	turbo, turbine
Tus	Tuscaloosa
TV	television
TVA	temporary voluntary allowance

TVD	true vertical depth
TVP	true vapor process
TW	tank wagon
Tw Cr	Twin Creek
TWI	techniques of water-resources investigations
twp	township
TWR	tower
twst	townsite
twst off	twisted off
TWTM	too wet (weak) to measure
TWX	teletype
ty	type
typ	typical
tywr	typewriter

U

U	unclassified
U/	upper (i.e., U/Simpson)
U/C	under construction
U/L	upper and lower
U/W	used with
U/WTR	utility water
UC	upper casing
UCH	use customer's hose
UD	under digging
UFD	utility flow diagram
UG	under gauge
UG	underground
UGL	universal gear lubricant
UHF	ultra-high frequency
ULJ	perforating, Ultrajet
ult	ultimate
UM	Umiat Meridian (Alaska)
UMB	umbrella (s)
un	unit
UNBAL	unbalanced

unbr	unbranded
UNC	unified coarse thread
unconf	unconformity
uncons	unconsolidated
undiff	undifferentiated
unf	unfinished
UNF	unified fine thread
uni	uniform
Univ	university, universal
UNLD	unloading
UNLDR	unloader
UOCO	Union Oil Company
UPS	uninteruptible power supply
UR	underreaming
UR	unsulfonated residue
UR	used rod
USG	United States gauge
UST	ultrasonic test
UT	ultrasonic examination
UT	upper tubing
UT	upthrown
UTL	utility
UTM	universal transverse mercator
UV	ultraviolet
UV	Union Valley
Uvig. lir.	Uvigerina lirettensis

V

V	valve
V	viscosity
v, V	volt
V	volume
v.	very (as very tight)
v.n.	very noticeable

v.c.	very common
V.P.S.	very poor sample
v.r.	very rare
v-f-gr	very fine-grained
v-HOCM	very heavily oil-cut mud
v-sli	very slight
V/DWG	vendor drawing
V/L	vapor-liquid ratio
V/S	velocity survey
v%	volume-percent
va	Volt-ampere
vac	vacant
vac	vacation
vac	vacuum
Vag. reg.	Vaginuline regina
Val	Valera
Vang	Vanguard
vap (r)	vapor (izor)
var	variable, various
var	volt-ampere reactive
vari	variegated
VARN	varnish
VCP	vitrified clay pipe
vel	velocity
vent	ventilator
Ver Cl	Vermillion Cliff
Verd	Verdigris
vert	vertical
ves	vesicular
VESS	vessel
vfg	very fine-grain (ed)
VGC	viscosity-gravity constant
VHF	very high frequency
VHGCM	very heavily (highly) gas-cut mud
VHGCSW	very heavily (highly) gas-cut salt water
VHGCW	very heavily (highly) gas-cut water
VHO&GCM	very heavily (highly) oil and gas-cut mud

VHO&GCSW	very heavily (highly) oil and gas-cut salt water
VHO&GCW	very heavily (highly) oil and gas-cut water
VHOCM	very heavily (highly) oil-cut mud
VHOCSW	very heavily (highly) oil-cut salt water
VHOCW	very heavily (highly) oil-cut water
Vi	Viola
VI	viscosity index
VIB	vibrate (tor) (ing)
Virg	Virgelle
vis	viscosity
vis	visible
vit	vitreous
Vks	Vicksburg
VLAC	very light amber cut
vlv	valve
VM&P Naphtha	varnish makers and painters naphtha
VOC	volatile organic compounds
Vogts	Vogtsberger
vol	volume
vol%	volume-percent
vol. eff.	volumetric efficiency
VOLT	voltage
VP	vapor pressure
VPU	vapor recovery unit
VR	vapor recovery
vrs	varas
vrtb	vertebrate
vrtl	vertical
vrvd	varved
VS	velocity survey
vs	versus
VSC	volumetric subcommittee
VSGCM	very slightly gas-cut mud
VSGCSW	very slightly gas-cut salt water

VSGCW	very slightly gas-cut water
VSM	vertical support member
VSO&GCM	very slightly oil and gas-cut mud
VSO&GCSW	very slightly oil and gas-cut salt water
VSOCM	very slightly oil-cut mud
VSOCSW	very slightly oil-cut salt water
VSOCW	very slightly oil-cut water
VSP	very slightly porous
VSSG	very slight show of gas
VSSO	very slight show of oil
vt	vapor temperature
vug	vuggy
vug	vugular

—————— **W** ——————

W	wall (if used with pipe)
W	water-supply paper
w	watt
W	west
W	wide
W Cr	Wall Creek
w shd	washed
W.O.B.	weight on bit
W-F	Washita-Fredericksburg
w-hr	watt-hour
W&R	wash and ream
w/	with
W/CLR	water cooler
W/L	water load
W/O	west offset
W/O	without
W/SSO	water with slight show of oil
W/sulf O	water with sulfur odor
W/2	west half

w%	weight-percent
Wab	Wabaunsee
WAB	weak air blow
WACT	weight averaged catalyst temperature
Wad	Waddell
WAG	water-alternating gas (or water and gas)
Wap	Wapanucka
War	Warsaw
Was	Wasatch
WaSd	Waltersburg sand
Wash	Washita
WB	water blanket
WB	wet bulb
WB	Woodbine
WBIH	went back in hole
WC	water closet
WC	water cushion (DST)
WC	water cut
WC	wildcat
WC	Wolfe City
WCM	water-cut mud
WCO	water-cut oil
WCTS	water cushion to surface
WD	water depth
WD	water disposal well
WD	wiring diagram
Wd R	Wind River
Wdfd	Woodford
WE	weld ends
Web	Weber
Well	Wellington
WF	new field wildcat, dry
WF	waterflood
WF	wide flange
WFD	new field wildcat, discovery
WFD	wildcat field, discovery
wgt	weight
WH	wellhead

Wh Dol	white dolomite
Wh Sd	white sand
WHIP	wellhead injection pressure
whip	whipstock
WHL	wheel
whse	warehouse
whsle	wholesale
wht	white
WI	washing in
WI	water injection
WI	working interest
WI	wrought iron
Wich Alb	Wichita Albany
Wich.	Wichita
WIH	water in hole
Willb	Willberne
Win	Winona
Winf	Winfield
Wing	Wingate
Winn	Winnipeg
WIP	work in place
WIW	water injection well
wk	weak
wk	week
wkd	worked
wkg	working
wko	workover
wkor	workover rig
WL	water loss
WL	well lines
WL	west line
WL	wireline
wlbr	wellbore
WLC	wireline coring
wld	welded, welding
WLD/DET	welding detail (s)
wldr	welder
WLT	wireline test
WLTD	wireline total depth
W'ly	westerly

WN	weld neck
WNSO	water not shut off
WO	waiting on
WO	wash oil
WO	wash over
wo	washout
WO	wildcat outpost, dry
WO	work order
WO	workover
WOA	waiting on acid
WOA	waiting on allowable
WOB	waiting on battery
WOC	wating on cement
WOCR	waiting on completion rig
WOCT	waiting on cable tools or completion tools
WODP	waiting on drillpipe
WOE	successful wildcat outpost
WOG	waiting on geologist
WOG	water oil or gas
Wolfc	Wolfcamp
WOO	waiting on orders
Wood	Woodside
Woodf	Woodford
WOP	waiting on permit
WOP	waiting on pipe
WOP	waiting on plastic
WOP	waiting on pump
WOPE	waiting on production equipment
WOPL	waiting on pipeline
WOPT	waiting on potential test
WOPU	waiting on pumping unit
WOR	waiting on rig or rotary
WOR	water-oil ratio
WORT	waiting on rotary tools
WOS	washover string
WOSP	waiting on state potential
WOST	waiting on standard tools
WOT	waiting on test or tools

WOT&C	waiting on tank and connection
WOW	waiting on weather
WP	new pool wildcat, dry
WP	wash pipe
WP	well pad
WP	working pressure
WPD	new pool wildcat, discovery
WPM	well pad manifolding
wpr	wrapper
WPT	Windfall Profit Tax
WR	White River
Wref	Wreford
WRG	wiring
WRTB	wash and ream to bottom
WS	shallower pool wildcat, dry
WS	water saturation
WS	whipstock
WS	worldscale
WSD	shallower pool wildcat, discovery
WSD	whipstock depth
wsh (g)	wash (ing)
WSIM	water separation index modified
WSO	water shutoff
WSONG	water shutoff no good
WSOOK	water shutoff OK
WST	waste
WST	water source wells
WSW	water supply well
WT	wall thickness (pipe)
wt	weight
wt%	weight-percent
WTB	wash to bottom
wtg	waiting
WTH/PRF	weatherproof
wthr(d)	weather (ed)
wtr (y)	water, watery
wtr. cush	water cushion

WTR/PRF	waterproof
WTR/T	watertight
WTS	water to surface
WUT	water-up to
WW	wash water
WW	water well
Wx	Welex
Wx	Wilcox

X

X	salt
X-bdd(ing)	crossbedded, crossbedding
X-hvy	extra heavy
X-line	extreme line (casing)
X-over	crossover
X-R	X-ray
X-REF	cross reference
X-SECT	cross section
x-stg	extra strong
x/n	crystalline
XFMR	transformer
XHGR	extra-heavy grade pipe
Xing	crossing
Xlam	cross-laminated
Xln	crystalline
XMTR	transmitter
XO	crossover
XO-sub	crossover sub
xtal	crystal
Xtree	Christmas tree
XW	salt water
XX-Hvy	double extra heavy
XX-STR	double extra strong

————————— **Y** —————————

Y	Yates
yd	yard (s)
yel	yellow
YIL	yellow indicating lamp
YMD	your message of date
YMY	your message yesterday
Yoak	Yoakum
YP	yield point
yr	year
Yz	Yazoo

————————— **Z** —————————

zen	zenith
Zil	Zilpha
ZN	zinc
Zn	zone

MISCELLANEOUS

10^2	trillion
12 GA W.W.S.	12 gauge wire-wrapped screen (in a liner)
3 PH	three phase
3P ST SW	triple pole single throw switch
3P SW	triple pole switch
4P ST SW	triple pole switch
4P SW	four pole switch
8rd	eight round pipe

/ft	per foot
/L	line, as in E/L (east line)
%	percent
°API	degrees, API
°C	degrees Centrigrade, degrees Celsius
°F	degrees Fahrenheit
μg	microgram (s)

A

abandoned	abd
abandoned, salvage deferred	**ASD**
abandoned gas well	**abd-gw**
abandoned location	**abd loc**
abandoned oil & gas well	**abd-ogw**
abandoned oil well	**abd-ow**
about	**abt**
above	**abv**
abrasive jet	**abrsi jet**
absolute	**abs**
absolute bottom-hole location	**ABHL**
absolute open flow potential (gas well)	**AOF**
absorber	**absr**
absorption	**absrn**
abstract	**abst**
abstract (i.e., A-10)	**A**
abundant	**abun**
accelerometers	**ACCEL**
access	**ACC**
accessory	**ACCESS**
account (ing)	**acct**
accounts receivable	**A/R**
accumulative, accumulator	**accum**
acid	**ac**
acid frac	**AF**
acid fracture treatment	**acfr**
acid residue	**AR**
acid-soluble oil	**ASO**
acid treat (ment)	**AT**
acid water	**AW**
acid-cut mud	**ACM**

acid-cut water	**ACW**
acidity or alkalinity	**pH**
acidize (ed) (ing)	**acd**
acidized with	**A/**
acoustic caliper	**CMA**
acoustic cement	**A-Cem**
acre-feet	**ac-ft**
acre (s)	**ac**
acreage	**ac, acrg**
acrylonitrile butadiene styrene rubber	**ABS**
actual	**ACT**
actual drilling	**AD**
actual drilling cost	**ADC**
actual drilling time	**ADT**
actual jetting time	**AJT**
actuated, actuator	**ACT**
adapter	**adpt**
addition or modification request	**AMR**
additional	**addl**
additive	**add**
adhesive	**ADH**
adjustable	**adj**
adjustable spring wedge	**ASW**
adjustments and allowances	**A&A**
administration, administrative	**adm**
adomite	**ADOM**
adsorption	**adspn**
advanced	**advan**
aeration, aerator	**AER**
affidavit	**afft**
affirmed	**affd**
after acidizing	**AA**
after condenser	**AF/COND**
after cooler	**AF/CLR**
after federal income tax	**AFIT**
after fracture	**AF**
after receipt of order (purchasing term)	**ARO**
after shot	**AS**

after top center	**ATC**
after treatment	**AT**
agglomerate	**aglm**
aggregate	**AGGR**
agitator	**AG**
air conditioning	**A/C**
air cooled	**A/CLD**
air cooler	**A/CLR**
air quality control region	**AQCR**
air quality maintenance area	**AQMA**
alarm	**alm**
Alaskan North Slope	**ANS**
Albany	**Alb**
alcoholic	**alc**
algae	**alg**
alignment (ing)	**ALIGN**
alkaline, alkalinity	**alk**
alkalinity or acidity	**pH**
alkylate, alkylation	**alkyl**
all thread	**AT**
allocation	**ALOC**
allowable not yet available	**ANYA**
allowable, allowance	**ALLOW**
alloy	**ALY**
along	**alg**
alternate	**alt**
alternating current	**AC**
altitude	**ALT**
aluminum	**AL**
aluminum conductor steel reinforced	**ACSR**
ambient	**amb**
American Chemical Society	**ACS**
American melting point	**AMP**
American Petroleum Institute	**API**
American Public Health Association	**APHA**
American Society for Testing & Materials	**ASTM**
American Standards Association	**ASA**

American Steel & Wire gauge	**AS&W ga**
American Water Works Association	**AWWA**
American Wire gauge	**AWG**
ammeter	**AMM**
ammonia	**NH₃**
ammonium chloride	**NH₄Cl**
amorphous	**amor**
amortization	**amort**
amount	**amt**
amount not reported	**ANR**
ampere	**amp**
ampere-hour	**amp-hr**
amphipore	**amph**
amphistegina	**amph**
analysis, analytical	**anal**
anchor (age)	**ANC**
and husband	**et con.**
and husband	**et vir.**
and others	**et al.**
and the following	**et seq.**
and wife	**et ux.**
angle, angular	**ang**
angstrom unit	**Å**
angulogerina	**angul**
anhydrite stringer	**AS**
anhydrite, anhydritic	**anhy**
anhydrous	**anhyd**
annubar	**ANUB**
annular velocity	**AV**
annulus	**an**
annulus	**Ann**
annunciator	**ANUC**
apartment	**apt**
apparatus	**APPAR**
apparent (ly)	**apr**
appears, appearance	**app**
appliance	**appl**
application	**applic**
applied	**appl**

approved	**appd**
approved total depth	**ATD**
approximate (ly)	**approx**
aqueous	**aq**
aragonite	**arag**
Arapahoe	**Ara**
Arbuckle	**Arb**
Archeozoic	**Archeo**
architectural	**arch**
area of mutual interest	**AMI**
arenaceous	**aren**
argillaceous	**arg**
argillite	**arg**
Arkadelphia	**Arka**
arkose (ic)	**ark**
armature	**arm**
aromatics	**arom**
around	**arnd**
arrange (ed) (ing) (ment)	**ARR**
articles published in outside journals/books	**OP**
artificial lift	**AL**
as soon as possible	**ASAP**
asbestos	**asb**
Ashern	**Ash**
asphalt, asphaltic	**asph**
asphaltic stain	**astn**
assembly	**assy**
assigned	**assgd**
assignment	**asgmt**
assistant	**asst**
associate (ed) (s)	**assoc**
association	**assn**
Association of American Railroads	**AAR**
Association of Official Agricultural Chemists	**AOAC**
at rate of	**ARO**
Atlas Bradford modified	**ABM**
atmosphere, atmospheric	**atm**

Atoka	**At**
atomic	**at**
atomic weight	**at wt**
attach (ed) (ing) (ment)	**ATT**
attempt (ed)	**att**
attorney	**atty**
Audit Bureau of Circulation	**ABC**
auditorium	**aud**
Austin	**Aus**
Austin chalk	**AC**
Authorization for Commitment	**AFC**
Authorization for Expenditure	**AFE**
Authorization to Proceed	**ATP**
authorized	**auth**
authorized depth	**AD**
Authorized for Construction	**AFC**
automatic	**auto**
automatic custody transfer	**ACT**
automatic data processing	**ADP**
automatic transmission fluid	**ATF**
automatic volume control	**AVC**
automotive	**auto**
automotive gasoline	**autogas**
Aux Vases sand	**AV**
auxiliary	**aux**
auxiliary flow diagram	**AFD**
available	**avail**
average	**avg**
average flowing pressure	**AFP**
average freight rate assessment	**AFRA**
average injection rate	**AIR**
average penetration rate	**APR**
average treating pressure	**ATP**
average tubing pressure	**ATP**
aviation	**av**
aviation gasoline	**avgas**
awaiting	**awtg**
award	**AWD**
azeotropic	**aztrop**
azimuth	**az**

B

back flush	**BKFLSH**
back pressure	**BP**
back-pressure valve	**BPV**
back scuttled	**B/S**
back to back	**B/B**
backed out (off)	**BO**
backwash	**BKWSH**
baffle	**BFL**
bailed	**bld**
bailed dry	**B/dry**
bailer	**blr**
bailer feed water	**BFW**
bailing	**blg**
balance	**BAL**
ball joint	**B/JT**
ball sealers	**BS**
ball valve	**B/Vlv**
Balltown sand	**Ball**
band (ed)	**bnd**
barge deck to mean low water	**BD-MLW**
barite (ic)	**bar**
Barker Creek	**Bark Crk**
Barlow lime	**Bar**
barometer, barometric	**bar**
barrel	**bbl**
barrel water load	**BWL**
barrels acid	**BA**
barrels acid residue	**BAR**
barrels acid water	**BAW**
barrels acid water per day	**BAWPD**
barrels acid water per hour	**BAWPH**
barrels acid water under load	**BAWUL**
barrels condensate	**BC**
barrels condensate per day	**BCPD**
barrels condensate per hour	**BCPH**
barrels condensate per million	**BCPMM**
barrels diesel oil	**BDO**
barrels distillate	**BD**

barrels distillate per day	**BDPD**
barrels distillate per hour	**BDPH**
barrels fluid	**BF**
barrels fluid per day	**BFPD**
barrels fluid per hour	**BFPH**
barrels formation water	**BFW**
barrels frac oil	**BFO**
barrels fresh water	**BFW**
barrels liquid per day	**BLPD**
barrels load	**BL**
barrels load & acid water	**BL&AW**
barrels load condensate	**BLC**
barrels load condensate per day	**BLCPD**
barrels load condensate per hour	**BLCPH**
barrels load oil	**BLO**
barrels load oil per day	**BLOPD**
barrels load oil per hour	**BLOPH**
barrels load oil recovered	**BLOR**
barrels load oil to be recovered	**BLOTBR**
barrels load oil yet to recover	**BLOYTR**
barrels load water	**BLW**
barrels load water per day	**BLWPD**
barrels load water per hour	**BLWPH**
barrels load water to recover	**BLWTR**
barrels mud	**BM**
barrels new oil	**BNO**
barrels new water	**BNW**
barrels oil	**BO**
barrels oil per calendar day	**BOPCD**
barrels oil per day	**BOPD**
barrels oil per hour	**BOPH**
barrels oil per producing day	**BOPPD**
barrels per barrel	**B/B**
barrels per day	**B/D**
barrels per hour	**B/hr**
barrels per minute	**B/M**
barrels per stream day	**BPSD**
barrels per stream day (refinery)	**B/SD**
barrels per well per day	**BPWPD**

barrels pipeline oil	**BPLO**
barrels pipeline oil per day	**BPLOPD**
barrels salt water	**BSW**
barrels salt water per day	**BSWPD**
barrels salt water per hour	**BSWPH**
barrels water	**BW**
barrels water overload	**BWOL**
barrels water per day	**BOPD**
barrels water per hour	**BWPD**
Bartlesville	**Bart**
basal	**bsl**
Basal Oil Creek sand	**BOCS**
base	**B/**
Base Blane	**B. Bl**
base of the salt	**B slt**
Base Pennsylvanian	**BP**
base plate	**BSPL**
base salt	**B/S**
basement	**bsmt**
basement (granite)	**base**
basic sediment	**BS**
basic sediment & water	**BS&W**
basket	**bskt**
Bateman	**Bate**
battery	**btry**
Baume	**Be**
beaded and center beaded	**B & CB**
Bear River	**Bear Riv**
bearing	**brg**
Bearpaw	**BP**
Beckwith	**Beck**
becoming	**bec**
bedding	**BDNG**
before acid treatment	**BAT**
before federal income tax	**BFIT**
before top dead center	**BTDC**
Beldon	**Bel**
belemnites	**Belm**
bell and bell	**B & B**
bell and flange	**B & F**
bell and spigot	**B & S**

Belle City	**Bel C**
Belle Fourche	**Bel F**
benchmark	**BM**
bending schedule	**B/S**
Benoist (Bethel) sand	**Ben, BT**
bent & bowed pipe	**B&B**
Benton	**Ben**
bentonite	**Bent**
benzene	**bnz**
benzene toluenexylene (unit)	**BTX (unit)**
Berea	**Be**
between	**btw**
bevel (ed)	**bev**
bevel both ends	**BBE**
bevel large end	**BLE**
bevel one end	**BOE**
bevel small end	**BSE**
beveled for welding	**BV/WLD**
beveled end	**B.E.**
bid summary	**BID SUM**
Big Horn	**B. Hn.**
Big Injun	**B. Inj.**
Big Lime	**B. Ls**
Bigenerina	**Big.**
Bigenerina floridana	**Big. f.**
Bigenerina humblei	**Big. h.**
Bigenerina nodosaria	**Big. nod.**
bill of lading	**B/L**
bill of material	**B/M**
bill of sale	**B/S**
billion	**B**
billion cubic feet	**BCF, Bcf**
billion cubic feet per day	**BCFD, Bcfd**
billion standard cubic feet	**Bscf**
billion standard cubic feet per day	**Bscf/D, Bscfd**
binary	**BIN**
biochemical oxygen demand	**BOD**

biotite	**bio**
Birmingham (or Stubbs) iron wire gauge	**BW ga**
Birmingham wire gauge	**Bwg**
bitumen	**bit**
bituminous	**bit**
black	**blk**
Black Leaf	**Blk Lf**
Black Lime	**Blk Li**
Black Magic (mud)	**BM**
black malleable iron	**BMI**
Black River	**B. Riv**
black sulfur water	**BSUW**
blank liner	**blk lnr**
blast cabinet	**Bl/Cb**
blast joint	**BL/JT**
bleeding	**bldg**
bleeding gas	**bldg**
bleeding oil	**bldo**
blend (ed) (er) (ing)	**BLND**
blew out	**BO**
blind flange	**BLD FLG, BF**
Blinebry	**Blin**
block	**blk**
block valve	**BV**
blocked off	**BO**
Blossom	**Blos**
blow	**blo**
blowdown	**BLDWN**
blow-down test	**BDT**
blower	**BLWR**
blowout equipment	**BOE**
blowout preventer	**BOP**
blowout preventer equipment	**BOPE**
blue	**bl**
board	**bd**
board-foot; board-feet	**bd ft**
Bodcaw	**Bod**
body wall loss	**BWL**

boiled water	**BW**
boiler	**BLR**
boiler feed water	**BFW**
boiling point	**BP**
Bois d'Arc	**Bd'A**
Bolivarensis	**Bol.**
Bolivina a.	**Bol. a.**
Bolivina floridana	**Bol. flor.**
Bolivina perca	**Bol. p.**
Bone Spring	**BS**
Bonneterre	**Bonne**
booster	**BSTR**
borehole compensated sonic	**BHCS**
bottom sediment & water	**BS&W**
bottom (ed)	**btm (d)**
bottom choke	**btm chk**
bottom hole	**BH**
bottom-hole assembly	**BHA**
bottom-hole choke	**BHC**
bottom-hole flowing pressure	**BHFP**
bottom-hole location	**BHL**
bottom-hole money	**BHM**
bottom-hole orientation	**BHO**
bottom-hole pressure	**BHP**
bottom-hole pressure bomb	**BHPB**
bottom-hole pressure, closed (See SIBHP and BHSIP)	**BHPC**
bottom-hole pressure, flowing	**BHPF**
bottom-hole pressure survey	**BHPS**
bottom-hole shutin pressure	**BHSIP**
bottom-hole temperature	**BHT**
bottom of given formation (i.e., B/Frio)	**B/**
bottom sediment	**BS**
bottom settlings	**BS**
bottoms sediment and water	**B.S.&W.**
boulders	**bldrs**
boundary	**bndry**
box end	**be**
box (es)	**bx**

brace (ed) (ing)	**BRC**
brachiopod	**brach**
bracket (s)	**brkt(s)**
brackish (water)	**brksh**
Bradenhead flange	**BHF**
brake horsepower	**bhp**
brake horsepower-hour	**bhp-hr**
brake mean effective pressure	**BMEP**
brake specific fuel consumption	**BSFC**
brakes	**BRKS**
break (broke)	**brk**
breakdown	**bkdn**
breakdown acid	**BDA**
breakdown pressure	**BDP**
breaker	**BRKR**
breccia	**brec**
bridge plug	**BP**
bridged back	**BB**
Bridger	**Brid**
Brinell hardness number	**BHN**
British Standards Institution	**BSI**
British thermal unit	**BTU**
brittle	**brtl, brit**
broke (break) down formation	**BDF**
broken	**brkn**
broken sand	**brkn sd**
bromide	**brom**
brown	**brn or br**
Brown and Sharpe gauge	**B&S ga**
Brown lime	**Brn Li**
brown oil stain	**BOS**
brown shale	**brn sh**
brownish	**bnish**
bryozoa	**bry**
Buckner	**Buck**
buck off	**b/off**
buck on	**b/on**
Buckrange	**Buckr**
budgeted depth	**BD**
buff	**bf**

building	bldg
building derrick	bldg drk
building rig	BR
building roads	bldg rds
Buliminella textularia	Bul. text.
bulk plant	BP
bulk vessel	B/VESS
bull plug	BP
bullets	blts
Bullwaggon	Bull W
bumper	bmpr
bundle	BDL
Burgess	Burg
burner	bunr
bushel	bu
bushing	BSHG
butane and propane mix	BP mix
butane-butene fraction	BB fraction
butt weld	BW, BTWLD
butterfly valve	BRFL/V, BTFL/V
buttress thread	butt
buttress thread coupling	BTC
buzzer	BUZ
bypass	BYP
bypass cooler	BP/CLR

C

cabinet	CAB
cable (ing)	CBL
cable tool measurement	CTM
cable tools	CT
Caddell	Cadd
cadmium plate	CD PL
cake	ck

calcareous, calcerenite	**calc**
calceneous	**cale**
calcite, calcitic	**cal**
calcium	**calc**
calcium-base grease	**calc gr**
calcium chloride	**CaCl$_2$**
calcium oxide	**CaO**
calculate (ed), calculation	**Calc**
calculated absolute open flow	**CAOF**
calculated open flow (potential)	**COF**
calendar day	**CD**
calibrate (tion)	**CALIBR**
caliche	**cal**
California Coordinate System	**CCS**
caliper log	**CAL**
caliper survey	**cal**
caulking	**CLKG**
calorie	**cal**
Calvin	**Calv**
Cambrian	**Camb**
Camp Colorado	**Cp Colo**
Cane River	**Cane Riv**
canvas-lined metal petal basket	**CLMP**
canyon	**cany, cyn**
Canyon Creek	**Cany Crk**
capacity, capacitor	**cap**
Capitan	**Cap**
carbon copy	**CC**
carbon dioxide	**CO$_2$**
carbon disulfide	**CS$_2$**
carbon monoxide	**CO**
carbon oxygen	**CO**
carbon residue (Conradson)	**CR Con**
carbon steel	**CS**
carbon tetrachloride	**carb tet**
carbonaceous	**carb**
carburetor air temperature	**CAT**
care of	**c/o**
Carlile	**Car**
carload	**CL**

Carmel	**Carm**
Carrizo	**Cz**
carton	**ctn**
cased hole	**C/H**
cased reservoir analysis	**CRA**
casing	**csg**
casing cemented (depth)	**CC**
casing choke	**Cck**
casing collar locator	**CCL**
casing collar perforating record	**CCPR**
casing flange	**CF**
casing point	**csg pt, CP**
casing pressure	**csg press, CP**
casing pressure, shut in	**CPSI**
casing pressure, closed	**CPC**
casing pressure, flowing	**CPF**
casing seat	**CS**
casing set at	**CSA**
casinghead	**csg hd**
casinghead flange	**CHF**
casinghead gas	**CHG**
casinghead pressure	**CHP**
Casper	**Casp**
cast carbon steel	**CCS**
cast iron	**CI**
cast steel	**CS**
cast-iron bridge plug	**CIBP**
cat-cracked light gas oil	**CCLGO**
Cat Creek	**Cat Crk**
Catahoula	**Cat**
catalog	**CAT**
catalyst, catalytic	**CAT**
catalytic cracker	**Cat ckr**
catalytic cracking unit	**CCU**
cathode ray tube	**CRT**
cathodic	**cath**
Cattleman	**Ctlmn**
caustic	**caus**
caving (s)	**cvg(s)**

cavity	**cav**
Cedar Mountain	**Cdr Mtn**
cellar	**cell**
cellar & pits	**C & P**
cellular	**cell**
Celsius	**C**
cement (ed)	**cem**
cement (ed) (ing)	**cmt (d) (g)**
cement dump bailer	**CDB**
cement evaluation	**CET**
cement friction reducer	**CFR**
cement friction retarder	**CFR**
cement in place	**CIP**
cement to surface	**CTS**
cemented through perforations	**cp's**
cementer	**cmtr**
Cenozoic	**Ceno**
center (ed)	**cntr**
center (land description)	**C**
center line	**C/L**
center of casinghead flange	**CCHF**
center of gravity	**CG**
center of tubing flange	**CTF**
center section line	**CSL**
center to center	**C to C**
center to end	**C to E**
center to face	**C to F**
Centigrade	**C**
centigram	**cg**
centiliter	**cl**
centimeter	**cm**
centimeter-gram-second system	**cgs**
centimeters per second	**cm/sec**
centipoise	**cp**
centistokes	**cs**
central compressor plant	**CCP**
centralizers	**cent**
centrifugal	**centr**
centrifuge	**cntf**
cephalopod	**ceph**

Ceratobulimina eximia	**Cert. ex.**
certified	**CERT**
certified drawing outline	**CDO**
certified public accountant	**CPA**
cetane number	**CN**
chain operated	**CH OP**
chairman	**chrm**
chalcedony	**chal**
chalk	**chk**
chalky	**chky**
chamber	**CHMBR**
chamfer	**CHAM**
change (ed) (ing)	**chng**
changed (ing) bits	**CB**
changed drillpipe	**chngd DP**
channel	**CHNL**
Chappel	**Chapp**
characteristics	**CHAR**
charge (ed) (ing)	**chrg (d) (ing)**
Charles	**Char**
chart	**cht**
Chattanooga shale	**Chatt**
check	**ck**
check valve	**CHKV**
checked	**chkd**
checkerboard	**Chkbd**
checkered plate	**CHKD PL**
chemical oxygen demand	**COD**
chemical products	**chem prod**
chemical, chemist, chemistry	**chem**
chemically pure	**cp**
chemically retarded acid	**CRA**
Cherokee	**Cher**
chert	**cht**
cherty	**chty**
Chester	**Ches**
chicksan	**cksn**
Chimney Hill	**Chim H**
Chimney Rock	**Chim R**

Chinle	**Chin**
chitin (ous)	**chit**
chloride (s)	**chl**
chlorinator	**CHLR**
chlorine	**CL₂**
chlorine log	**chl log**
chloritic	**chl**
choke	**chk**
Chouteau lime	**Chou**
Christmas tree	**Xtree**
chromatograph	**chromat**
chrome molybdenum	**cr moly**
chromium	**chrome**
Chugwater	**Chug**
Cibicides	**Cib.**
Cibicides hazzardi	**Cib. h.**
Cimarron	**Cima**
circle	**cir**
circuit	**cir**
circular	**cir**
circular mils	**cir mils**
circulate & condition	**C&C**
circulate and reciprocate	**C&R**
circulate bottoms up	**CBU, ccBU**
circulate (ing) (tion)	**circ**
circulated and condition mud	**C&CM**
circulated and conditioned hole	**C&CH**
circulated out	**CO**
circumference	**CRCMF**
Cisco	**Cis**
Clagget	**Clag**
Claiborne	**Claib**
clarifier	**CLFR**
Clarksville	**Clarks**
class	**CL**
classification	**CLASS**
clastic	**clas**
Clavalinoides	**Clav**
clay filled	**CF**

Claystone	clyst
Clayton	Clay
Claytonville	Clay
clean out	CO
clean out & shoot company	CO & S
clean up	CU
clean (ed) (ing)	cln (d) (g)
cleaned out to total depth	COTD
cleaning out, cleaned out	CO
cleaning to pits	CTP
clear, clearance	clr
Clearfork	Clfk
clearing	clrg
Cleveland	Cleve
Cleveland open cup	COC
Cliff House	Cliff H
clockwise	cw
closed	clsd
closed cup	CC
closed hole	CH
closed hydrocarbon drain	CHD
closed-in pressure	CIP
Cloverly	Clov
coarse crystalline	crs-xln
coarse grained	cse gr, cg
coarse (ly)	crs, c
coat and wrap (pipe)	C & W
coated	ctd
Cockfield	Cf
Coconino	Coco
Codell	Cod
Cody (Wyoming)	Cdy
coefficient	coef
coiled tubing unit	CTU
coke oven gas	COG
cold drawn	CD
cold finished	CF
cold rolled	CR
cold rolled steel	CRS
cold water equivalent	CWE

cold working pressure	**CWP**
Coleman Junction	**Cole Jct**
collar	**colr**
collect (ed) (ing) (tion)	**coll**
collector	**CLTR**
Color American Standard Test Method	**Col ASTM**
colored	**COL**
column	**COL**
Comanche	**Com**
Comanche Peak	**Com Pk**
Comanchean	**Cmchn**
Comatula	**Com**
combined, combination	**comb**
combustion	**COMB**
coming out of hole	**COOH**
commenced	**comm**
comment	**COMT**
commercial	**coml**
commission	**comm**
commission agent	**C/A**
commissioner	**commr**
common	**com**
Common Business-Oriented Language	**COBOL**
common data base	**CDB**
communication	**comm**
community	**comm**
compact	**cmpt**
companion flange bolt & gasket	**CFB & G**
companion flange one end	**CFOE**
companion flanges bolted on	**CFBO**
company	**Co**
company operated	**Co. Op.**
company-operated service stations	**Co. Op. S.S.**
compartment	**compt**
compensated neutron log	**CNL**
complete with	**C/W**
computer-processed interpretation	**MCT**

cubic feet per minute	**cu ft/min, CFM**
cubic feet per pound	**CFP**
cubic feet per second	**cu ft/ sec, CFS**
cubic foot	**cu ft**
cubic inch	**cu in.**
cubic meter	**cu m**
cubic meters per day	**m³/d**
cubic yard	**cu yd**
cubical	**CUB**
culvert	**culv**
cumulative	**cum**
Curtis	**Cur**
curve	**CRV**
cushion	**cush**
customer	**CUST**
cut across grain	**CAG**
Cut Bank	**Cut B**
cut drilling line	**CDL**
cutbank	**cutbk**
Cutler	**Cutl**
cutting oil	**Cut Oil**
cutting oil soluble	**Cut Oil Sol**
cutting oil-active-sulfurized dark	**Cut Oil Act Sul-dk**
cutting oil-active-sulfurized transparent	**Cut Oil Act Sul-transpt**
cutting oil-inactive-sulfurized	**Cut Oil Inact Sul**
cutting oil-straight mineral	**Cut Oil St Mrl**
cuttings	**ctg(s)**
Cyclamina	**Cyc.**
Cyclamina cancellata	**Cyc. canc.**
cycles per minute	**cpm**
cycles per second	**cps**
cyclone	**CYC**
cylinder	**cyl**

Cypress sand	**Cy Sd**
cypridopsis	**cyp.**

D

daily allowable	**DA**
daily average injection barrels	**DAIB**
Dakota	**Dak**
damper	**dmpr**
Dantzler	**Dan**
dark	**dk**
dark brown oil	**DBO**
dark brown oil stains	**DBOS**
Darwin	**Dar**
data processing	**DP**
data sheet	**D/S**
date actually started drilling	**spud date**
date of first production	**DFP**
datum	**dat, DM**
datum faulted out	**DFO**
Davit	**DVT**
day	**D**
day to day	**D/D**
days since spudded	**DSS**
dead	**dd**
dead oil show	**DOS**
deadweight tester	**DWT**
deadweight tons	**DWT**
Deadwood	**Deadw**
deaerator	**deaer**
dealer	**dlr**
dealer tank wagon	**DTW**
deasphalting	**deasph**
debutanizer	**debutzr**
decibel	**db**
decigram	**dg**
deciliter	**dl**

decimal	**dec**
decimeter	**dm**
decline	**decl**
decrease (ed) (ing)	**decr**
deep pool test	**DPT**
deepen	**dpn**
deepening	**dpg**
deethanizer	**deethzr**
deflection	**defl**
Degonia	**Deg**
degree day	**DD**
degree (s)	**deg**
degree API	**°API**
degrees Centigrade	**°C**
degrees Fahrenheit	**°F**
deisobutanizer	**deisobut**
Del Rio	**Del R**
Delaware	**Dela**
Delaware River Area Petroleum Refineries	**DRAPR**
delayed coker	**DC**
delivery (ed) (ability)	**delv**
delivery point	**delv pt**
demand meter	**DM**
demolition	**dml**
demurrage	**demur**
dendrite (ic)	**dend**
dense	**ds, dns**
density log	**D/L, DENL**
department	**dept**
Department of Energy	**DOE**
depletion	**depl**
depreciation	**deprec**
depropanizer	**deprop**
depth	**dpt**
depth bracket allowable	**DBA**
depth recorder	**dpt rec**
derrick	**drk**
derrick floor	**DF**
derrick floor elevation	**DFE**
Des Moines	**Des M**

desalter	**desalt**
description	**desc**
Desert Creek	**Des Crk**
design	**dsgn**
Desk and Derrick	**D & D**
desorbent	**desorb**
destination	**dstn**
desulfurizer	**desulf**
desuperheater	**DSUPHTR**
detail (s)	**det**
detector	**det, DCTR**
detergent	**deterg**
detrital	**detr, dtr**
develop	**DVL**
develop (ed) (ment)	**devel**
development	**D**
development gas well	**DG**
development oil	**DO**
development oil well	**DO**
development redrill (sidetrack)	**DR**
development well, carbon dioxide	**DC**
development well, helium	**DH**
development well, sulfur	**DSU**
development well workover	**DX**
deviate, deviation	**dev**
deviation degrees	**DD**
Devonian	**Dev**
dew point	**DP**
dewatering	**DWTR**
dewaxing	**dewax**
Dexter	**Dext**
diagonal	**diag**
diagram	**diag**
diameter	**dia**
diamond bit	**DB**
diamond core	**DC**
diamond core bit	**DCB**
diaphragm	**diaph**
dichloride	**dichlor**
dichloro-diphenyl-trichloroethane	**DDT**

diesel (oil)	**dsl**
diesel fuel	**DF**
diesel hydrogen desulfurization	**DHDS**
diesel index	**D.I.**
Diesel No. 2	**D-2**
diesel oil cement	**DOC**
diethanolamine	**DEA**
diethanolamine unit	**DEA unit**
diethylene	**diethy**
different (ial) (ence)	**diff**
differential pressure	**D/P**
differential valve (cementing)	**DV**
digging cellar	**DC**
digging cellar and slush pits	**DCLSP**
digging slush pits	**DSP**
digital	**DGTL**
diglycolamine	**DGA**
diluted	**dilut**
dimension	**dim**
dimethyl sulfide	**DMS**
diminish (ing)	**dim**
Dinwoody	**Din**
dipmeter	**DM**
direct (tion) (tor)	**dir**
direct current	**DC**
directional drilling	**dir drlg**
directional survey	**dir sur, DS**
dirty water disposal	**DWD**
discharger	**disc, disch**
Discorbis	**Disc.**
Discorbis gravelli	**Disc. grav.**
Discorbis normada	**Disc. norm.**
Discorbis yeguaensis	**Disc. y.**
discount	**disc**
discover (y) (ed) (ing)	**disc**
discovery allowable requested	**DAR**
dismantle	**disman**
dismantle (ing)	**dsmt (g)**
displaced, displacement	**displ**
disseminated	**dism**
distance	**dist**

distillate	**dstl**
distillate-cut mud	**DCM**
distillate, distillation	**dist**
distribute (ed) (ing) (ion)	**distr**
distributed control system	**DCS**
district	**dist**
ditto	**do**
division	**div**
division office	**D/O**
division order	**D.O.**
dock operating building	**DOBLDG**
Dockum	**Doc**
doctor-treating	**doc-tr**
document	**doc**
documentation	**DOCREQ**
dogleg severity	**DLS**
doing business as	**d/b/a**
dolomite (ic)	**dolo**
dolstone	**dolst**
domestic	**dom**
domestic airline	**dom AL**
domestic water	**DOM WTR**
Dornick Hills	**Dorn H**
Dothan	**Doth**
double	**DBL**
double end	**DE**
double extra heavy	**XX-Hvy**
double extra strong	**XX-STR**
double hub	**DH**
double pole (switch)	**DP**
double pole double base (switch)	**DPDB**
double pole double throw switch	**DPDTSW**
double pole single base (switch)	**DPSB**
double pole single throw switch	**DPST SW**
double pole switch	**DP SW**
double random lengths	**DRL**
Douglas	**Doug**
down	**dn**
downthrown	**DT**

dozen	**doz**
draft gauge	**DG**
drain	**dr**
drainage	**drng**
drawing	**DWG**
drawworks	**dwks**
dressed dimension four sides	**d-d-4-s**
dressed dimension one side and one edge	**d-d-1-s-1-e**
dressed four sides	**d-4-s**
dressed one side	**d-1-s**
dressed two sides	**d-2-s**
Dresser Atlas	**DA**
drier, drying	**dry**
drift angle	**DA**
drill	**drl**
drill (ed) (ing) out	**DO**
drill (ed) (ing) plug	**D/P**
drill and complete	**D & C**
drill collar	**DC**
drill floor	**DF**
drillpipe	**DP**
drillpipe measurement	**DPM**
drillpipe unloaded	**DPU**
drill site	**DS**
drillsite facility	**DSF**
drillstem	**DS**
drillstem test	**DST**
drilled	**drld**
drilled-out cement	**DOC**
drilled-out depth	**DOD**
drilled-out plug	**DOP**
driller	**drlr**
driller's top	**D/T**
driller's total depth	**DTD**
drilling	**drlg**
drilling and well completion	**DWC**
drilling break	**DB**
drilling deeper	**DD**
drilling line	**DL**

drilling mud	**DM**
drilling suspended indefinitely	**DSI**
drilling tender	**D/T**
drilling time	**DT**
drilling with air	**DWA**
drilling with gas	**DWG**
drilling with mud	**DWM**
drilling with oil	**DWO**
drilling with salt water	**DWSW**
drillstem test with straddle packers	**DST (Strd)**
drive (ing) (er)	**DRV (R), dr**
dropped	**dropd**
drum	**dr, DRM**
druse	**dr**
drusy	**drsy**
dry and abandoned	**D & A**
Dry Creek	**Dr Crk**
dry desiccant dehydrator	**DDD**
dry film thickness	**DFT**
dry gas	**DG**
dry-hole contribution	**DHC**
dry hole drilled deeper	**DHDD**
dry-hole money	**DHM**
dry hole reentered	**DHR**
dual	**D**
dual (double) wall packer	**DWP**
dual injection focus log	**DIFL**
dual lower casing	**DLC**
dual lower tubing	**DLT**
dual upper casing	**DUC**
dual upper tubing	**DUT**
dually completed	**DC**
Duck Creek	**Dk Crk**
dumped	**DMPD**
Dun & Bradstreet	**D & B**
Duperow	**Dup**
duplex	**dx**
duplicate	**dup**

duration	**DUR**
Dutcher	**Dutch**
dyna-drilling	**DD**
dynamic	**dyn**

— E —

each	**ea**
Eagle	**Egl**
Eagle Ford	**EF**
Eagle Mills	**EM**
Earlsboro	**Earls**
early well tie-ins	**EWT**
east	**E**
east boundary line	**E/BL**
East Cimarron Meridian (Oklahoma)	**ECM**
east half	**E/2**
east line	**E/L**
east of Rockies	**EOR**
east of west line	**E of W/L**
east offset	**E/O**
east quarter	**E/4**
easterly	**E'ly**
Eau Claire	**Eau Clr**
eccentric	**ECC**
echinoid	**ech**
economics, (y), economizer	**econ**
Ector (County TX)	**Ect**
education (tor)	**Educ (r)**
Edwards	**Edw**
Edwards lime	**Ed lm**
effective	**eff**
effective depth	**E.D.**
effective horsepower	**EHP**
efficiency	**eff**
effluent	**effl**
eight round pipe	**8rd**

ejector	**eject**
Elbert	**Elb**
elbow (s)	**ell(s), ELB**
electric accounting machines	**EAM**
electric log tops	**EL/T**
electric resistance weld	**ERW**
electric weld	**EW**
electric (al)	**elec**
electromagnetic	**EMN, ELEC/MAG**
electromagnetic induction	**EMNI**
electromotive force	**EMF**
electron volts	**ev**
electronic data processing	**EDP**
element, elementary	**elem**
elevation (height)	**EL**
elevation ground	**el gr**
elevation, elevator	**elev**
Elgin	**Elg**
eliminate (tor) (ed)	**ELIM**
Ellenburger	**Ellen**
Ellis-Madison contact	**EMS**
Elmont	**Elm**
Embar	**Emb**
emergency	**emer**
emergency order	**EO**
emergency shutdown	**ESD**
employee	**empl**
empty container	**MT**
emulsion	**emul**
enamel	**enml**
enclosure	**encl**
end of file	**EOF**
end of line	**EOL**
end of month	**EOM**
end of quarter	**EOQ**
end of year	**EOY**
end point	**EP**
end to end	**E/E**
Endicott	**End**

endothyra	**endo**
engine	**eng**
engineer (ing)	**engr (g)**
Englewood	**Eglwd**
enhanced oil recovery	**EOR**
enlarged	**enl**
Entrada	**Ent**
entrance	**ENT**
entry	**ent**
envelope	**ENV**
environment	**ENVIR**
environmental assessment	**EA**
environmental impact report	**EIR**
environmental impact statement	**EIS**
Eocene	**Eoc**
Eponides	**Ep.**
Eponides yeguaensis	**Ep. y.**
equal, equalizer	**eq**
equation (before a number)	**Eq.**
equilibrium flash vaporization	**EFV**
equipment	**equip**
equivalent	**equiv**
erection	**erect**
erection mark	**ERC/MK**
Ericson	**Eric**
estate	**est**
estimate (ed) (ing)	**est**
estimated time of arrival	**ETA**
estimated total depth	**ETD**
estimated ultimate recovery	**EUR**
estimated yearly consumption	**EYC**
ethane	**eth**
ethylene	**ethyle**
ethylene dichloride	**EDC**
euhedral	**euhed**
European melting point	**EMP**
Eutaw	**EU**
evaluate	**eval**
evaporation, evaporate	**evap**
even-sorted	**ev-sort**
examination	**EXAM**

example	**EX**
excavation	**exc**
excellent	**excl**
except	**ex**
exchanger	**exch**
excitation	**EXC**
executive	**EXEC**
executor	**Exr**
executrix	**Exrx**
Exeter	**Ex**
exhaust	**exh**
exhibit	**exh**
existing	**exist**
expansion	**exp**
expansion joint	**EXP JT**
expected date of delivery	**EDD**
expendable plug	**exp plg**
expense	**exp**
expire (ed) (ing) (ation)	**expir**
exploratory	**E**
exploratory well	**EW**
exploratory, exploration	**expl**
explosion proof	**EX-PRF**
explosive	**explos**
extended, extension	**ext (n)**
extension manhole	**Ext M/H**
exterior	**extr**
external	**ext**
external upset end	**EUE**
extra heavy	**X-hvy**
extra heavy grade pipe	**XHGR**
extra strong	**x-stg**
extraction	**extrac**
extreme line (casing)	**X-line**
extreme pressure	**EP**

F

fabricate (ed) (tion)	**fab**
face of stud	**FOS**

face to face	**F to F**
faced and drilled	**F & D**
facet (ed)	**fac**
facility (ies)	**FACIL**
facility capacity limits	**FCL**
failed	**fld**
failure	**fail**
faint	**fnt**
faint air blow	**FAB**
fair	**fr**
Fall River	**Fall Riv**
Farmington	**Farm**
farmout	**FO**
farmout option	**F/O opt**
fasten (ing) (er)	**FSTN**
fault	**flt**
faulted out	**FO**
fauna	**fau**
favosites	**fvst**
federal	**fed**
Federal Employers Liability Act	**FELA**
Federation of Sewage and Industrial Wastes Association	**FSIWA**
feed	**FD**
feed effluent	**FD EFF**
feed rate	**FR**
feed water	**FD/WTR**
feeder	**fdr**
feedstock	**FS**
feet, foot	**ft**
feet per hour	**ft/hr**
feet per minute	**fmp**
feet per minute	**ft/min**
feet per second	**fps**
feet per second	**ft/sec**
feldspar	**fld**
female	**FEM**
female pipe thread	**FPT**
female to female angle	**FFA**
female to female globe (valve)	**FFG**

Ferguson	**Ferg**
ferric sulfate	**Fe$_2$(SO$_4$)$_3$**
ferruginous	**ferr**
Ferry Lake anhydrite	**FLA**
fertilizer	**fert**
fiberglass-reinforced plastic	**FRP**
fibrous	**fib**
field	**fld**
field authorized to commence operations	**FACO**
field fabricated	**F/FAB**
field fuel gas unit	**FFGU**
field pressure test flow diagram	**FPTFD**
field purchase order	**FPO**
field receiving report	**FRR**
field wildcat	**FWC**
figure	**fig**
fill up	**FU**
fillet weld	**FW**
filter	**FLTR**
filter cake	**FC**
filtrate	**filt**
final	**fin**
final (flowing) tubing pressure	**FTP**
final boiling point	**FBP**
final bottom-hole pressure, flowing	**FBHPF**
final bottom hole pressure, shut in	**FBHPSI**
Final Environmental Impact Statement	**FEIS**
final flowing pressure	**FFP**
final fluid level	**FFL**
final hydrostatic pressure	**FHP**
final open	**FO**
final pressure	**FP**
final report for rig	**FRR**
final report for well	**FRW**
final shutin pressure	**FSIP**
finding of no significant impact	**FONSI**
fine	**fn**

fine grained	**f-gr**
finely	**fnly**
finely crystalline	**f/xln**
finish	**FNSH**
finish all over	**FAO**
finish going in hole	**FGIH**
finish going in with	**FGIW**
finish grade	**FIN GR**
finish (ed)	**fin**
finished drilling	**fin drlg**
finished in hole	**FIH**
fire water	**F/WTR**
fire-resistant oil	**F-R oil**
fireproof	**fprf**
fiscal year ending	**FYE**
fishing	**fish, fsg**
fishing for	**FF**
fissile	**fisl**
fissure	**fis**
fittings	**ftg**
fixed	**fxd**
fixed carbon	**FC**
fixture	**fix**
flaky	**flk**
flammable	**FLMB**
flammable liquid building	**FL/BD**
flange, (ed) (es)	**flg (d) (s)**
flanged and dished (heads)	**F & D**
flanged and spigot	**F & S**
flanged and screwed	**F&S**
flanged gate valve	**FGVV**
flanged one end, welded one end	**FOE-WOE**
flash point, Cleveland Open Cup	**Fl-COC**
flashing	**FL**
flat face	**FF**
flathead	**Flath**
flattened	**flat**
flexible	**flex**
flexibox	**FLXBX**

Flippen	**Flip**
float	**flt**
float collar	**FC**
float shoe	**FS**
flotation	**fltn**
floating	**fltg**
floor	**FL**
floor drain	**FD**
Florence Flint	**Flor Fl**
flow	**flo**
flow (ed), (ing)	**flw (d) (g)**
flow control valve	**FCV**
flow diagram	**F/DIA, FD, F-DIA**
flow-indicating controller	**FIC**
flow-indicating ratio controller	**FIRC**
flow indicator	**FI**
flow line	**FL**
flowmeter	**F-MET**
flow rate	**FR**
flow recorder	**FR**
flow recorder control	**FRC**
flow sheet	**F-SHT**
flow station	**FS**
flow switch	**F/SW**
flowed (ing) at a rate of	**FARO**
flowed, flowing	**F/, fl/**
Flowerpot	**Flwrpt**
flowing	**flg**
flowing bottom hole pressure	**FBHP**
flowing by heads	**FBH**
flowing casing pressure	**FCP**
flowing on test	**FOT**
flowing pressure	**Flwg Pr, FP**
flowing surface pressure	**FSP**
flue	**flu**
fluid	**flu**
fluid catalytic cracking	**FCC**
fluid in hole	**FIH**
fluid level	**FL**

fluid to surface	**FTS**
fluorescence, fluorescent	**fluor**
flush	**FL**
flush joint	**FJ**
flushed	**flshd**
flushing oil	**FLO**
focused log	**FOCL**
foliated	**fol**
foot-candle	**ft-c**
foot-pound	**ft-lb**
foot-pound per hour	**ft-lb/hr**
foot-pound-second (system)	**f-p-s**
footing, footage	**ftg**
for example	**e.g.**
for your information	**FYI**
Foraker	**Forak**
foraminifera	**foram**
foreman	**f'man**
forge (ed) (ing)	**FRG**
forged steel	**FST, FS**
formation	**fm**
formation density	**FD**
formation density correlated	**FDC**
formation density log	**FDL**
formation gas-oil ratio	**F/GOR**
formation interval tester	**FIT**
formation test	**FT**
formation water	**Fm W**
Fort Chadborne	**Ft C**
Fort Hayes	**Ft H**
Fort Riley	**Ft R**
Fort Union	**Ft U**
Fort Worth	**Ft W**
Fortura	**Fort**
forward	**fwd**
fossiliferous	**foss**
foundation	**fdn**
Fountain	**Fount**
four pole single throw switch	**4P ST SW**
four pole switch	**4P SW**

four-wheel drive	**FWD**
Fox Hills	**Fox H**
frac finder (log)	**FF**
fractional	**fr**
fractionation, fractionator, fractional	**fract**
fracture gradient	**F.G.**
fracture, fractured, fractures	**frac (d) (s)**
fragment	**frag**
frame, framing	**FRM (G)**
framework	**frwk**
franchise	**fran**
Franconia	**Franc**
Fredericksburg	**Fred**
Fredonia	**Fred**
free on board	**FOB**
free point back off	**FPBO**
free point indicator	**FPI**
freezer	**frzr**
freezing point	**FP**
freight	**frt**
frequency	**freq**
frequency meter	**FM**
frequency modulation	**FM**
fresh	**frs**
fresh break	**FB**
fresh water	**FW**
fresh water	**fwtr**
friable	**fri**
friction-reducing agent	**FRA**
froggy	**Frgy**
from	**fr**
from east line	**FE/L, FEL, fr E/L**
from north line	**FN/L, FNL, fr N/L**
from northeast line	**FNEL**
from northwest line	**FNWL**
from south and west lines	**FS&WLs**

from south line	**FS/L, FSL, fr S/L**
from southeast line	**FSEL**
from southwest line	**FSWL**
from west line	**FW/L, FWL, fr W/L**
front	**fr**
front & side	**F/S**
frontier	**fron**
frosted	**fros, fr**
frosted quartz grains	**FQG**
Fruitland	**Fruit**
fuel gas	**FG**
fuel oil	**FO**
fuel oil equivalent	**F.O.E.**
fuel oil return	**FOR**
fuel oil supply	**FOS**
fuels & fractionation	**F & F**
fuels & lubricants	**F & L**
full freight allowed (purchasing term)	**FFA**
full hole	**FH**
full length drift	**FLD**
full of fluid	**FF**
full open head	**FOH**
full opening	**FO**
Fullerton	**Full**
functional check out	**FCO**
funnel viscosity	**FV**
furfural	**furf**
furnance fuel oil	**FFO**
furnance	**furn**
furnish (ed)	**FURN**
furniture and fixtures	**furn & fix**
Fuson	**Fus**
Fusselman	**fussel**
fusulinid	**fusul**
future	**fut**

G

gauge (ed) (ing)	**ga**
Galena	**Glna**
Gallatin	**Gall**
galled threads	**gld thd**
gallon (s)	**gal**
gallons acid	**GA**
gallons breakdown acid	**GBDA**
gallons condensate per day	**GCPD**
gallons condensate per hour	**GCPH**
gallons gelled water	**GGW**
gallons heavy oil	**GHO**
gallons mud acid	**GMA**
gallons of oil per day	**GOPD**
gallons of oil per hour	**GOPH**
gallons of solution	**gal sol**
gallons of water per hour	**GWPH**
gallons oil	**GO**
gallons per day	**GPD**
gallons per hour	**GPH**
gallons per minute	**gal/min, GPM**
gallons per second	**GPS**
gallons per thousand cubic feet	**gal/Mcf**
gallons regular acid	**GRA**
gallons salt water	**GSW**
gallons water	**GW**
galvanized	**galv**
gamma ray	**GR**
gamma ray spectroscopy tool	**GST**
gas	**G**
gas & mud-cut oil	**G&MCO**
gas and oil	**G&O**
gas and oil-cut mud	**G&OCM**
gas cap participating area	**GCPA**
gas in pipe	**GIP**
gas·injection	**GI**
gas injection well	**GIW**

gas lift	**GL**
gas lift gas distribution	**GGD**
gas lift oil	**GLO**
gas lift transfer	**GLT**
gas-liquid ratio	**GLR**
gas odor	**GO**
gas odor distillate taste	**GODT**
gas pay	**GP**
gas purchase contract	**GPC**
gas reserve group	**GRG**
gas rock	**G Rk**
gas sales contract	**GSC**
gas show	**GS**
gas to surface	**GTS**
gas to surface (time)	**GTS**
gas too small to measure	**GTSTM**
gas unit	**GU**
gas volume	**GV**
gas volume not measured	**GVNM**
gas well	**GW**
gas well shut in	**GSI**
gas-condensate ratio	**GCR**
gas-cut	**GC**
gas-cut acid water	**GCAW**
gas-cut distillate	**GCD**
gas-cut load oil	**GCLO**
gas-cut load water	**GCLW**
gas-cut mud	**GCM**
gas-cut oil	**GCO**
gas-cut salt water	**GCSW**
gas-cut water	**GCW**
gas-distillate ratio	**GDR**
gas-fluid ratio	**GFR**
gas-handling study group	**GHSG**
gas-oil contact	**GOC**
gas-oil ratio	**G/O, GOR**
gas-water contact	**GWC**
gas-well gas	**GWG**
gaseous nitrogen	**G-N$_2$**
gaseous oxygen	**G-O$_2$**
gasket	**gskt**

gasoline	**gaso**
gasoline plant	**GP**
gastropod	**gast**
gathering line	**G/L**
gauge	**gge**
gauge ring	**GR**
gelled	**gel**
general	**genl**
general arrangement	**GA**
General Electric Company	**GE**
General Land Office (Texas)	**GLO**
General Motors Corporation	**GM**
generation, generator	**gen**
geological quadrangle map	**GQM**
geology (ist) (ical)	**geol**
geophysical investigation map	**GPM**
geophysics (ical)	**geop**
geopressure development, failure	**PD**
geopressure development, success	**PDS**
Georgetown	**Geo**
geothermal	**geo, GT**
geothermal development, failure	**GD**
geothermal development, success	**GDS**
geothermal wildcat, failure	**GW**
geothermal wildcat, success	**GWD**
Gibson	**Gib**
Gilcrease	**Gilc**
gilsonite	**gil**
glass, glassy	**gls**
glauconite, glauconitic	**glau**
Glen Dean lime	**GD Li**
Glen Rose	**GR**
Glenwood	**Glen**
globe valve	**GLBVV**
Globigerina	**Glob**
Glorieta	**Glor**
glycol	**glyc**
gneiss	**gns**
going in hole	**GIH**

Golconda lime	**Gol Li**
good	**gd**
good fluorescence	**GFLU**
good odor & taste	**gd o&t**
good show of gas	**GSG**
good show oil	**GSO**
good show oil and gas	**GSO&G**
Goodland	**Gdld,** **Good L**
Goodwin	**Gdwn**
goose egg	**G egg**
Gorham	**Gor**
Gouldbusk	**Gouldb**
government	**govt**
governor	**gov**
grade	**gr**
grading	**grdg**
grading location	**grdg loc**
gradiomanometer	**GRAD**
gradual, gradually	**grad**
grain	**gr**
grained (as in fine grained)	**gnd**
grains per gallon	**gg, GPG**
gram	**g, gm**
gram molecular weight	**g mole**
gram-calorie	**gm-cal**
Graneros	**Granos**
Granite Point Field	**GPF**
granite wash	**gran w**
granite	**gran**
grant (of land)	**grt**
granular	**grnlr**
graptolite	**grap**
grating	**grtg**
gravel	**gvl**
gravel packed	**GVLPK**
gravitometer	**grvt**
gravity	**grav, gr,** **GTY,** **GRVTY**

gravity °API	**gr °API**
gravity meter	**G.M.**
gray	**gry**
Gray sand	**Gr Sd**
Grayburg	**Grayb**
Grayson	**Gray**
graywacke	**gywk**
grease	**gr**
greasy	**gsy**
green	**grn**
Green River	**Grn Riv**
green royalty	**gr roy**
green shale	**grn sh**
Greenhorn	**GH**
grind out	**GO**
gritty	**grty**
grooved	**grv**
grooved ends	**GE**
gross	**grs**
gross acre-feet	**GAF**
gross weight	**gr wt**
ground	**gr, grnd**
ground level	**GL**
ground measurement (elevation)	**GM**
group	**GRP**
guard log	**GRDL**
guidance continuance tool	**GCT**
guide shoe	**GS**
Gulf Research and Development Company	**GR&DC**
Gull River	**G. Riv**
gummy	**gmy**
gun barrel	**GB**
gun perforate	**G/P**
Gunsite	**Guns**
gusset	**GUS**
gypsiferous	**gypy**
gypsum	**gyp**
Gyroidina	**Gyr.**
Gyroidina scal	**Gyr. sc.**

H

Hackberry	**Hackb**
hackly	**hky**
hand hole	**HH**
hand-control valve	**HCV**
handle	**hdl**
handling	**HNDLG**
handwheel	**HND/WHL**
hanger	**hgr**
Haragan	**Hara**
harbor	**hbr**
hard	**hd**
hard lime	**hd li**
hard sand	**hd sd**
Hardinsburg sand (local)	**Hburg**
hardness	**hdns**
hardware	**hdwe**
Haskell	**Hask**
Haynesville	**Haynes**
hazardous	**haz**
head	**hd**
header	**hdr**
headquarters	**HQ**
heat exchanger	**HX, HTX**
heat tracing (ed)	**HT**
heat-treated alloy	**HTA**
heat-treated, heater treater	**HT**
heater	**htr**
heating and ventilating	**H & V**
heating oil	**HO**
heating ventilating and air conditioning	**HVAC**
heavily	**hvly**
heavily (highly) gas-cut salt water	**HGCSW**
heavily (highly) gas-cut water	**HGCW**
heavily (highly) oil-cut mud	**HOCM**
heavily (highly) oil-cut salt water	**HOCSW**
heavily (highly) oil-cut water	**HOCW**

heavily (highly) water-cut mud	**HWCM**
heavily gas-cut mud	**HGCM**
heavily oil and gas-cut mud	**HO&GCM**
heavy	**hvy**
heavy coker gas oil	**HCGO**
heavy cycle oil	**HCO**
heavy duty	**HD**
heavy fuel oil	**HFO**
heavy gas oil	**HVGO**
heavy hydrocrackate	**HUX**
heavy oil	**HO**
heavy reformate	**HR**
heavy steel drum	**HSD**
Heebner	**Heeb**
height	**hgt**
heirs	**hrs**
held by production	**HBP**
hematite	**hem**
Herington	**Her**
Hermosa	**Herm**
Hertz	**Hz**
heterostegina	**het**
hex head	**HEX HD**
hexagon (al)	**hex**
hexane	**hex**
Hickory	**Hick**
high detergent	**HD**
high gas-oil ratio	**HGOR**
high pressure	**HP**
high temperature	**HT**
high tension	**HT**
high viscosity	**HV**
high viscosity index	**HVI**
high voltage	**H-VOLT**
high volume lift	**HVL**
high-level shutdown	**HLSD**
high-pressure gas	**HPG**
high-pressure gauge	**HPG**
high-resolution dipmeter	**HRD**
high-temperature shutdown	**HTSD**
highly	**hily**

highway	hwy
Hilliard	Hill
hockleyensis	hock
Hogshooter	Hog
hold down	HLDN
hole full of oil	HFO
hole full of salt water	HFSW
hole full of sulfur water	HR Sul W
hole full of water	HFW
hole opener	HO, H.O.
holes per foot	HPF
Hollandberg	Holl
Home Creek	Home Cr
home office	HO
hook up	HU
hookwall packer	HWP
Hoover	Hov
hopper	hop
horizontal	horiz
horsepower	HP
horsepower-hour	hp-hr
Hospah	Hosp
hot dry rock development, failure	HD
hot dry rock development, successful	HDS
hot dry rock wildcat, failure	HW
hot dry rock wildcat, success	HWS
hot oil tar	HOT
hot-rolled steel	HRS
hour (s)	hr, HRS
house (ed) (ing)	HSE
house brand (regular grade of gasoline)	HB
Hoxbar	Hox
Humblei	Humb
Humphreys	Hump
hundred weight	cwt
Hunton	Hun
hydraulic	HYD
hydraulic pump	HP

Hydril	**HYD**
Hydril thread	**HYDT**
Hydril Type A joint	**HYDA**
Hydril Type CA joint	**HYDCA**
Hydril Type CS joint	**HYDCS**
hydro test	**HYDRO**
hydrocarbon	**HC**
hydrocarbon drain system	**HCDS**
hydrocracker	**H/C, HC**
hydrodesulfurizer	**HDS**
hydrofining	**hfg**
hydrogen	**H$_2$**
hydrogen delsulfurization	**HDS**
hydrogen ion concentration	**pH**
hydrogen sulfide	**H$_2$S**
hydrogenation	**HYGN**
Hydrologic Investigations Atlas	**HIA**
hydrostatic head	**HH**
hydrostatic pressure	**HP**
hydrostatic test	**HST**
hydrotreater	**hydtr**
hygiene	**hyg**
hyperbolic constant	**CSCH**
hyperbolic contangent	**COTH**
hyperbolic cosine	**COSH**
hypotenuse	**HYPO**
hyraulic horsepower	**H H P**

I

identification sign	**I.D. sign**
identify (ier) (ication)	**IDENT**
Idiomorpha	**Idio**
igneous	**ign**
ignition	**IGN**
illuminator (s)	**ILUM**
imbedded	**imbd**
immediate (ly)	**immed**

Imperial	**Imp**
Imperial gallon	**Imp gal**
impervious	**imperv**
impounding	**IMP**
impression block	**IB**
in hole	**IH**
inbedded	**inbded**
incandescent	**incd**
inch (es)	**in.**
inch-pound	**in.-lb**
inches mercury	**in. Hg**
inches per second	**in./sec**
incinerator, incineration	**INCIN**
include (ed) (ing)	**incl**
inclusions	**incls**
income (er) (ing)	**INCM**
incorporated	**Inc.**
increase (ed) (ing)	**incr**
increment	**incr**
indicate (s) (tion)	**indic**
indicated horsepower	**IHP**
indicated horsepower hour	**IHPHR**
indistinct	**indst**
individual	**indiv**
induction	**ind**
induction electrical survey	**IES**
indurated	**indr**
inflammable liquid	**Inf. L**
inflammable solid	**Inf. S**
inflow performance rate	**IPR**
information	**info**
infrared	**IR**
inhibitor	**inhib**
initial	**init**
initial air blow	**IAB**
initial boiling point	**IBP**
initial bottom-hole pressure	**IBHP**
initial bottom-hole pressure, flowing	**IBHPF**
initial bottom-hole pressure, shut in	**IBHPSI**
initial flowing pressure	**IFP**

initial fluid level	**IFL**
initial hydrostatic pressure	**IHP**
initial mud weight	**IMW**
initial open	**IO**
initial participating area	**IPA**
initial potential	**IP**
initial pressure	**IP**
initial production	**IP**
initial production flowed (ing)	**IPF**
initial production gas lift	**IPG**
initial production on intermitter	**IPI**
initial production plunger lift	**IPL**
initial production pumping	**IPP**
initial production swabbing	**IPS**
initial shutin tubing pressure	**ISITP**
initial shutin pressure (DST)	**ISIP**
initial vapor pressure	**IVP**
injection gas	**IG**
injection gas-oil ratio	**IGOR**
injection index	**II**
injection pressure	**Inj Pr**
injection rate	**IR**
injection water	**IW**
injection well	**IW**
injection, injected	**inj**
inland	**inl**
inlet	**inl**
inlet gate valve/inlet ball valve	**IGV/IBV**
Inoceramus	**Inoc**
input/output	**I/O**
inquire, inquiry	**INQ**
inside diameter	**ID**
inside screw (valve)	**IS**
insoluble	**insol**
inspect (ed) (ing) (tion)	**insp**
install (ed) (ing)	**inst**
install (ing) pumping equipment	**IPE**
installation operation & maintenance	**I-O-M**
installation (s)	**instl**

installing (ed) pumping equipment	**INPE**
instantaneous	inst
instantaneous shutin pressure (frac)	**ISIP**
institute	**inst**
instrument, instrumentation	**instr**
insulate	**insul**
insulate, insulation	**ins**
insurance	**ins**
integral	**INT**
integral joint	**IJ**
integrator	**intgr**
intention to drill	**ITD**
interbedded	**interbd**
interconnecting flow diagram	**I/CFD**
interconnection (ing)	**I/C, INTCON**
intercooler	**incolr, INCLR**
intercrystalline	**inter-xln**
interest	**int**
intergranular	**inter-gran, ingr**
interior	**int, INTR**
interlaminated	**inter-lam, intlam**
intermediate manifolds	**IMF**
intermediate pressure	**IP**
internal	**int, INTL**
internal flush	**IF**
Internal Revenue Service	**IRS**
internal upset ends	**IUE**
intersect	**ints**
intersection	**int**
interstitial	**intl**
interval	**intv**
intrusion	**intr**
invert (ed)	**inv**
invertebrate	**invrtb**

investment tax credit	**ITC**
invitation to bid	**ITB**
invoice	**inv**
Ireton	**Ire**
iridescent	**irid**
iron	**Fe**
iron body (valve)	**IB**
iron body brass (bronze) mounted (valve)	**IBBM**
iron body brass core (valve)	**IBBC**
iron case	**IC**
iron pipe size	**IPS**
iron pipe thread	**IPT**
ironstone	**Fe-st, irst**
irregular	**irreg**
isocracker	**ISO/CKR**
isolate (tor)	**ISOL**
isometric	**isom, ISO**
isopropyl alcohol	**IPA**
isothermal	**isoth**
issue	**ISS**
Iverson	**Ives**

J

jacket	**jac**
Jackson	**Jxn, Jack**
Jackson sand	**Jax sd**
jammed	**jmd**
Jasper (oid)	**Jasp**
Jefferson	**Jeff**
jelly-like colloidal suspension	**gel**
jet fuel (aviation)	**JFA**
jet perforated	**JP**
jet perforations per foot	**JP/ft**
jet propulsion fuel	**JP fuel**
jet shots per foot	**JSPF**
jet treating unit	**JTU**

job complete	**JC**
jobber	**jbr**
Joint Committee on Uniformity of Methods of Water Examination	**JCUMWE**
joint interest nonoperated (property)	**JINO**
joint operating agreement	**JOA**
joint operating provisions	**JOP**
joint operation	**J/O**
joint venture	**JV**
joint (s)	**jt(s)**
Jordan	**Jdn**
Joule	**J**
Judith River	**Jud Riv**
junction	**jct**
junction box	**JB**
junk (ed)	**jnk**
junk basket	**JB**
junk joint	**JJ**
junked and abandoned	**J&A**
Jurassic	**Jur**
jurisdiction	**juris**

--------------- **K** ---------------

Kaibab	**Kai**
Kansas City	**KC**
kaolin	**kao**
Kayenta	**Kay**
Keener	**Ke**
kelly bushing	**KB**
kelly bushing measurement	**KBM**
kelly drill bushing to landing flange	**KDB-LDG FLG**
kelly drill bushing to mean low water	**KDB-MLW**
kelly drill bushing to platform	**KDB-Plat**
kelly drive bushing	**KDB**

kelly drive bushing elevation	**KDBE**
Kelvin (temperature scale)	**K**
Keokuk-Burlington	**Keo-Bur**
kerosene	**kero**
ketone	**ket**
kettle	**KTLE**
Keystone	**Key**
Kiamichi	**Kia**
Kibbey	**Kib**
kick off	**KO**
kickoff point	**KOP**
kill (ed) well	**KW**
killed	**kld**
kiln dried	**KD**
kilocalorie	**kcal**
kilocycle	**kc**
kilogram	**kg**
kilogram-calorie	**kg-cal**
kilogram-meter	**kg-m**
kilohertz (see Hz-Hertz)	**KHz**
kiloliter	**kl**
kilometer	**km**
kilopascal	**kPa**
kilovar-hour	**kvar-hr**
kilovar, reactive kilovolt-ampere	**kvar**
kilovolt	**kv**
kilovolt peak	**kvp**
kilovolt-ampere	**kva**
kilovolt-ampere-hour	**kvah**
kilowatt	**kw**
kilowatt-hour	**kw-h**
kilowatt-hour meter	**kw-hm**
Kincaid lime	**Kin Li, KD**
Kinderhook	**Khk**
kinematic	**Kin**
kinematic viscosity	**KV**
Kirtland	**Kirt**
kitchen	**KIT**
KMA sand	**KMA**
knock down	**KD**

knock out	**KO**
known geothermal resource area	**KGRA**
Kootenai	**Koot**
Krider	**Kri**

--- **L** ---

La Motte	**La Mte**
labor	**lab**
laboratory	**lab**
ladder	**lad**
ladder & platform	**L & P**
lagging	**LAG**
laid (laying) down drill collars	**LDDCs**
laid (laying) down drillpipe	**LDDP**
laid down	**LD**
laid-down cost	**LDC**
Lakota	**Lak**
laminated, lamination (s)	**lam**
landing	**LDG**
land (s)	**ld (s)**
Landulina	**Land**
Lansing	**Lans**
lap joint	**LJ**
lapweld	**LW**
Laramie	**Lar**
large	**lrg, lg**
large-diameter flow line	**LDF**
Large Discorbis	**Lg Disc**
latitude	**lat**
Lauders	**Laud**
Layton	**Layt**
Le Comptom	**Le C**
leached	**lchd**
lead drill collar	**LDCX**
Leadville	**Leadv**
league	**Lge**
leak	**lk**

leakage	**LKG**
lease	**lse**
lease automatic custody transfer	**LACT**
lease crude	**LC**
lease operations	**LOS**
lease use (gas)	**L U**
Leavenworth	**Lvnwth**
left hand	**LH**
left in hole	**LIH**
legal subdivision (Canada)	**LSD**
length	**lg**
length overall	**LOA**
Lennep	**Len**
lense	**lns**
lenticular	**len**
less than truckload	**LTL**
less-than-carload lot	**LCL**
letter	**ltr**
level	**lvl**
level alarm	**LA**
level control valve	**LCV**
level controller	**LC**
level glass	**lg**
level indicator	**LI**
level indicator controller	**LIC**
level recorder	**LR**
level recorder controller	**LRC**
license	**lic**
Liebuscella	**Lieb**
light	**lt**
light barrel	**LB**
light brown oil stain	**LBOS**
light coker gas oil	**LCGO**
light fuel oil	**LFO**
light hydrocrackate	**LUX**
light iron barrel	**LIB**
light iron grease barrel	**LIGB**
light steel drum	**LSD**
lighting	**ltg**
lightning arrester	**LA**

lignite, lignitic	lig
lime, limestone	Li, lm, ls
limy	lmy
limy shale	Lmy sh
limit	lim
limited	ltd
line	LN
line pipe	L.P.
line pressure	LP
line, as in E/L (east line)	/L
linear	lin
linear foot	lin ft
liner	lnr, lin
linguloid	lngl
liquefaction	liqftn
liquefied petroleum gas	LP-Gas, LPG
liquid	liq
liquid level controller	LLC
liquid level gauge	LLG
liquid oxygen	LOX
liquid penetrant examination	LPE
liquid volume	LV
liquefied natural gas	LNG
list of components	LST/ COMPTS
liter	l
lithographic	litho
little	ltl
load	ld
load acid	LA
load oil	LO
load water	LW
loader	LDR
loading	LDG
loading dock	L-DK
local	LCL
local control panel	LCP
local injection plants	LIP
local purchase order	LPO
located, location	loc

location abandoned	**loc abnd**
location graded	**loc gr**
lock	**lk**
locker	**LKR**
lodge pole	**LP**
log mean temperature difference	**LMTD**
log total depth	**LTD**
logarithm (common)	**log**
logarithm (natural)	**ln**
long	**lg**
long coupling	**LC**
long handle/round point	**LH/RP**
long radius	**LR**
long-range automation plan	**LRAP**
long-range plan	**LRP**
long string	**LS**
long-term tubing test	**LTT**
long threads and coupling	**LT&C**
longitude (inal)	**long**
loop	**lp**
lost circulation	**LC**
lost circulation material	**LCM**
Lovell	**Lov**
Lovington	**Lov**
low pressure	**LP**
low viscosity index	**LVI**
low voltage	**L-VOLT**
low water loss	**LWL**
low-pressure separation	**LPS**
low-pressure separator	**LP sep**
low-temperature extraction unit	**LTX unit**
low-temperature separation unit	**LTS unit**
low-temperature shutdown	**LTSD**
lower	**lwr, low**
Lower Albany	**L/Alb**
lower anhydrite stringer	**LAS**
lower casing	**LC**
Lower Cretaceous	**L/Cret**
lower explosive limit	**LEL**

Lower Glen Dean	**LGD**
Lower Menard	**LMn**
lowermost flange	**LMF**
lower tubing	**LT**
Lower Tuscaloosa	**L/Tus**
lower, i.e., L/Gallup	**L/**
lube oil	**LO**
lubricate (ed) (ing) (tion)	**lub**
Lueders	**Lued**
lug cover type (5-gallon can)	**LC**
lug cover with pour spout	**LCP**
lumber	**lbr**
lumpy	**lmpy**
lustre	**lstr**

M

machine	**mach**
Mackhank	**Mack**
Madison	**Mad**
magnetic particle examination	**MPE**
magnetic, magnetometer	**mag**
magnetomotive force	**mmf**
main reaction furnace, "merf"	**MRF**
maintenance	**maint**
Manitoban	**Manit**
major, majority	**maj**
making	**MKG**
male and female (joint)	**M&F**
male pipe thread	**MPT**
male to female angle	**MFA**
malleable	**mall**
malleable iron	**MI**
management	**m'gmt**
manager	**mgr**
manhole	**MH**
manifold	**MF, man**
Manning	**Mann**

manual	**man, MNL**
manually operated	**man op**
manufacture (er)	**MFR**
manufactured	**mfd**
manufacturing	**mfg**
mapping subcommittee	**MSC**
Maquoketa	**Maq**
Marble Falls	**Mbl Fls**
macaroni tubing	**MT**
Marchand	**March**
marginal	**marg**
Marginulina	**Marg.**
Marginulina coco	**Marg. coco.**
Marginulina flat	**Marg. fl.**
Marginulina round	**Marg. rd.**
Marginulina texana	**Marg. tex.**
marine	**mar**
marine gasoline	**margas**
marine rig	**MR**
marine terminal	**M/T**
Marine Tuscaloosa	**M. Tus**
marine wholesale distributors	**MWD**
market demand factor	**MDF**
market (ing)	**mkt**
Markham	**Mark**
marking	**MRK**
marlstone	**mrlst**
marly	**Mly**
Marmaton	**Marm**
maroon	**mar**
Marshal (ling)	**MARSH**
masking plate	**M/PLT**
Massilina pratti	**Mass. pr.**
massive	**mass**
massive anhydrite	**MA**
master	**mstr**
master circuit board	**MCB**
master load list	**MLL**
material	**mat'l, mtl**

material safety data sheets	**MSDS**
material take-off	**MTO**
mathematics	**Math**
matrix	**Mtx**
matter	**mat**
maximum	**max**
maximum & final pressure	**M&FP**
maximum allowable operating pressure	**MAOP**
maximum allowable working pressure	**MAWP**
maximum casing pressure	**MCP**
maximum daily delivery obligation	**MDDO**
maximum efficient rate	**MER**
maximum flowing pressure	**MFP**
maximum operating pressure	**MOP**
maximum pressure	**MP**
maximum surface pressure	**MSP**
maximum top pressure	**MTP**
maximum total depth	**MTD**
maximum tubing pressure	**MTP**
maximum working pressure	**MWP**
Maywood	**May**
McClosky lime	**McC lm**
McCullough	**McCul**
McElroy	**McEl**
McKee	**McK**
McLish	**McL**
McMillan	**McMill**
Meakin	**Meak**
mean effective pressure	**MEP**
mean low water to platform	**MLW-PLAT**
mean low wave	**MLW**
mean sea level	**msl**
mean temperature difference	**MTD**
measure (ed) (ment)	**meas**
measure of hydrogen potential	**pH**
measured depth	**MD**
measured total depth	**MTD**

measuring & regulating station	**M & R Sta.**
mechanic (al), mechanism	**mech**
mechanical down time	**Mech DT**
mechanical flow diagram	**MFD**
mechanical properties	**MPL**
mechanism	**mchsm**
median	**med**
Medicine Bow	**Med B**
Medina	**Med**
medium	**med**
medium amber cut	**MAC**
medium fuel oil	**med FO**
medium grained	**m-gr, med gr, mg**
Medrano	**Medr**
Meeteetse	**Meet**
megacycle	**mc**
megahertz (megacycles per second)	**MHz**
megapascal	**mPa**
megawatt	**MW**
melting point	**MP**
member (geologic)	**mbr**
memo requesting quotes	**MRQ**
memorandum	**memo**
Menard lime	**Men li**
Menefee	**Mene**
Meramec	**Mer**
mercaptan	**mercap**
merchandise	**mdse**
mercury	**merc**
"merf", main reaction furnace	**MRF**
meridian	**merid**
Mesaverde	**Mvde**
mesh	**M**
Mesozoic	**Meso**
metal petal basket	**MPB**
metamorphic	**meta**
meter	**m, mtr**
meter run	**MR**

meter-kilogram	**m-kg**
methanator	**methr**
methane	**meth**
methane-rich gas	**MEG**
methanol	**methol**
methyl chloride	**meth-cl**
methylene blue	**meth-bl**
methylethylketon	**MEK**
methylisobutylketone	**MIK**
metric	**metr**
mezzanine	**mezz**
mhos per meter	**mho/m**
mica, micaceous	**mica**
micro-microfarad	**m-mf**
microampere	**m-a, ma**
microcrystalline	**micro-xin, micro-x**
microfarad	**mfd, m-f**
microfossil (iferous)	**micfos**
microgram	**m-g**
microgram (millionth)	**μg**
microinch	**m-in.**
micromicron	**m-m**
micron	**m, μ**
microsecond	**microsec, μsec**
microvolt	**μ-v, μv**
microwave	**MW**
middle	**M/, mdl, mid**
middle ground shoals	**MGS**
middle hydrocrackate	**MUX**
Midway	**Mwy**
mile (s)	**mi**
miles per hour	**MPH**
miles per year	**MPY**
military	**mil**
milky	**mky**
mill wrapped plain end	**MWPE**
milled	**mld**

milled one end	**M1E**
milled other end	**MOE**
milled two ends	**M2E**
milliampere	**ma**
millidarcies	**md**
milligram	**mg**
millihenry	**mh**
milliliter	**ml**
milliliters tetraethyl lead per gallon	**ml TEL/G**
millimeter	**mm**
millimeters of mercury	**mm Hg**
millimicron	**m**
milling .	**millg**
milling '	**mlg**
million (i.e., 9MM = 9,000,000)	**MM**
million British thermal units	**MMBTU**
million cubic feet	**MMcf**
million cubic feet per day	**MMcfd**
million electron volts	**MMev**
million reservoir barrels	**MMRVB**
million standard cubic feet per day	**MMscfd**
millions of barrels	**MMBLS**
milliotitic	**mill**
milliroentgen	**mr**
millisecond (s)	**ms**
millivolt	**mv**
mineral	**mnrl**
mineral interest	**MI**
minerals	**min**
Minerals Management Service	**MMS**
minimium	**min**
minimum pressure	**min P**
Minnekahta	**Mkta**
Minnelusa	**Minl**
minor	**MNR**
minute (s)	**min**
Miocene	**Mio**
miscellaneous	**misc**

Miscellaneous Field Studies Map	**MFSM**
Miscellaneous Investigations Series	**MIS**
miscible substance group	**MSG**
Misener	**Mise**
Mission Canyon	**Miss Cany**
Mississippian	**Miss**
mix and pump	**M&P**
mixed	**mxd**
mixer	**mix**
mixing	**MIXG**
mobile	**mob**
model	**mod**
moderate (ly)	**mod**
modification	**mod**
modular	**modu**
Moenkopi	**Moen**
moisture impurities and unsaponifiables (grease testing)	**MIU**
molar	**M**
molas	**mol**
mole	**mol**
molecular weight	**mol wt**
molecule, molecular	**MOL**
mollusca	**mol**
molybdenum	**MO**
Monel drill collars	**MDC**
monitor	**mon**
monoethanolamine	**MEA**
monthly volume operation plan	**MVOP**
Montoya	**Mont**
Moody's Branch	**MB**
Moore County line	**MC**
Mooringsport	**Moor**
more or less	**m/l**
Morrison	**Morr**
Morrow	**Mor**
mortgage	**mtge**
Mosby	**Mos**

motor	**MTR, mot**
motor control center	**MCC**
motor generator	**mg**
motor medium	**MM**
motor octane number	**MON**
motor oil	**MO**
motor oil units	**MOU**
motor severe	**MS**
motor vehicle fuel tax	**MVFT**
motor vehicle, motor vessel	**M/V**
mottled	**mott**
Mount Selman	**Mt. Selm**
mounted	**mtd**
mounting	**mtg**
mousehole	**MH**
moving	**movg**
moving (moved) in double drum unit	**MIDDU**
moving in (equipment)	**MI**
moving in and rigging up	**MIRU**
moving in cable tools	**MICT**
moving in completion rig	**MICR**
moving in completion unit	**MICU**
moving in equipment	**MIE**
moving in materials	**MIM**
moving in pulling unit	**MIPU**
moving in rig	**MIR**
moving in rigging up swabbing unit	**MIRUSU**
moving in rotary tools	**MIRT**
moving in service rig	**MISR**
moving in standard tools	**MIST**
moving in tools	**MIT**
moving out	**MO**
moving out (off) cable tools	**MOCT**
moving out (off) rotary tools	**MORT**
moving out completion unit	**MOCU**
moving out equipment	**MOE**
moving out rig	**MOR**
Mowry	**Mow**

Mt. Diablo	**MD**
mud acid	**MA**
mud acid wash	**MAW**
mud cake	**MC**
mud cleanout agent	**MCA**
mud cut	**MC**
mud filtrate	**MF**
mud logger	**ML**
mud logging unit	**MLU**
mud to surface	**MTS**
mud weight	**md wt, MW**
mud-cut acid	**MCA**
mud-cut gas	**MCG**
mud-cut oil	**MCO**
mud-cut salt water	**MCSW**
mud-cut water	**MCW**
mud/silt remover	**MSR**
muddy	**Mdy**
muddy salt water	**MSW**
muddy water	**MW**
mudstone	**mudst**
multigrade	**MG**
multiple service acid	**MSA**
multiply, multiplexer	**MULTX**
multipurpose	**MP**
multipurpose grease lithium base	**MPGH-lith**
multipurpose grease soap base	**MPGR-soap**
muscovite	**musc**

——————— **N** ———————

Nacotoch	**Nac**
nacreous	**nac**
nameplate	**NP**
naphfining unit	**NU**
naphtha	**nap**

naphtha-hydrogen desulfurization	**NHDS**
narrative	**NARR**
national	**nat'l**
national coarse thread	**NCT**
National Electric Code	**NEC**
National Fine (thread)	**NF**
National pipe thread	**NPT**
National pipe thread, female	**NPTF**
National pipe thread, male	**NPTM**
national pollution discharge elimination system	**NPDES**
natural	**nat**
natural flow	**NF**
natural gas	**NG**
natural gas liquids	**NGL**
nautical mile	**NMI**
Navajo	**Nav**
Navarro	**Navr**
Naval Petroleum Reserve	**NPR**
Naval Petroleum Reserve Alaska	**NPRA**
negative	**neg**
negligible	**neg**
negotiation	**NEGO**
neoprene	**npne**
net effective pay	**NEP**
net positive suction head	**NPSH**
net revenue interest	**NRI**
net tons	**NT**
neutral, neutralization	**neut**
neutralization number	**Neut. No.**
neutron lifetime log	**NLL**
New Albany shale	**New Alb**
new bit	**NB**
new field, discovery	**NFD**
new field wildcat	**NFW**
new field wildcat, discovery	**WFD**
new field wildcat, dry	**WF**
new oil	**NO**
new pool discovery	**NPD**

new pool exempt (non-operated)	**NPX**
new pool wildcat	**NPW**
new pool wildcat, discovery	**WPD**
new pool wildcat, dry	**WP**
new rod	**NR**
new source performance standards	**NSPS**
new total depth	**NTD**
Newburg	**Nbg**
Newcastle	**Newc**
Newton	**N**
Niagara	**Nig**
nickel plated	**NP**
Ninnescah	**Nine**
Niobrara	**Niob**
nipple	**nip, NPL**
nipple (ed) (ing) down blowout preventers	**NDBOPs**
nipple (ed) (ing) up blowout preventers	**NUBOPs**
nipple-down tree	**NDT**
nipple-up tree	**NUT**
nippled (ing) up	**NU, UP**
nippled down	**ND**
nippling up wellhead	**NUWH**
nitrogen	**N₂**
nitrogen blanket	**NB**
nitroglycerine	**nitro**
no appreciable gas	**NAG**
no change	**NC**
no core	**NC**
no fluid	**NF**
no fluorescence	**NF**
no fluorescence or cut	**NFOC**
no fuel	**NF**
no gas to surface	**NGTS**
no gauge	**NG**
no good	**NG**
no increase	**No Inc**

no order required	**NOR**
no paint on seams	**NPOS**
no production	**NP**
no recovery	**no rec, NR**
no report, not reported	**NR**
no show	**NS**
no show fluorescence or cut	**NSFOC**
no show gas	**NDG**
no show oil	**NSO**
no show oil and gas	**NSO&G**
no test	**N/tst**
no time	**NT**
no visible porosity	**NVP**
no water	**NW**
Noble-Olson	**NO**
Nodosaria blanpiedi	**Nod. blan.**
Nodosaria mexicana	**Nod. mex.**
nodule, nodular	**nod**
nominal	**nom**
nominal pipe size	**NPS**
noncontiguous tract	**NCT**
nondestructive testing	**NDT**
nondetergent	**ND**
nonemulsion acid	**NEA**
nonleaded gas	**NL gas**
nonoperated joint ventures	**NOJV**
nonoperating property	**NOP**
nonporous	**NP**
nonproducer	**N**
nonreturnable steel barrel	**NRSB**
nonreturnable steel drum	**NRSD**
nonreturnable, no returns, not reached	**NR**
nonrising stem (valve)	**NRS**
nonstandard	**nstd**
nonstandard service station	**N/S S/S**
nonupset	**NU**
nonupset ends	**NUE**
nonemulsifying agent	**NE**
nonflammable compressed gas	**nonf G**

Nonionella	**Non**
Nonionella cockfieldensis	**N. cock.**
Noodle Creek	**Ndl Cr**
normal	**nor**
normal (express concentration)	**N**
normally closed	**NC**
normally open	**NO**
north	**N**
north half	**N/2**
north line	**NL**
north offset	**N/O**
north quarter	**N/4**
northeast	**NE**
northeast corner	**NEC**
northeast line	**NEL**
northeast quarter	**NE/4**
northerly	**N'ly**
Northern Alberta Land Registration District	**NALRD**
northwest	**NW**
northwest corner	**NW/C**
northwest line	**NWL**
northwest quarter	**NW/4**
Northwest Territories	**NWT**
not applicable	**NA**
not available	**NA**
not completed	**NC**
not deep enough	**NDE**
not drilling	**ND**
not in contract	**NIC**
not on bottom	**NOB**
not prorated	**NP**
not pumping	**NP**
not suitable for coating	**NSC**
not to scale	**NTS**
not yet available	**NYA**
not yet drilled	**NYD**
Notary Public	**NP**
notice of intention to drill	**NID**
notice of violation	**NOV**

notice to proceed	**NTP**
nozzle	**noz**
Nuclear Regulatory Commission	**NRC**
nugget	**Nug**
number	**NO.**
number (before a number)	**No.**
numerous	**num**
nylon	**NYL**

O

Oakville	**Oakv**
object	**obj**
observation	**OBS**
obsolete	**obsol**
occasional (ly)	**occ**
ocean bottom suspension	**OBS**
octagon, octagonal	**oct**
octane	**oct**
octane number requirement	**ONR**
octane number requirement increase	**ONRI**
O'Dell	**Odel**
odor	**od**
odor, stain, and fluorescence	**OS&F**
odor, taste, & stain	**OT&S**
odor, taste, stain, & fluorescence	**OTS&F**
off bottom	**OB**
offshore	**offsh**
off/on location	**OL**
office, official	**off**
official potential test	**OPT**
offsite	**OFS**
O'Hara	**O'H**
Ohio River Valley Water Sanitation Commission	**ORSANCO**
ohm	**ohm**
ohm-centimeter	**ohm-cm**

ohm-meter	**ohm-m**
oil	**O**
oil abandoned well	**OAW**
oil and gas	**O&G**
oil and gas investigations chart	**OC-**
Oil & Gas Journal	**OGJ**
oil and gas lease	**O&GL**
oil and gas-cut acid water	**O&GCAW**
oil and gas-cut load water	**O&GCLW**
oil and gas-cut mud	**O&GCM**
oil and gas-cut salt water	**O&GCSW**
oil and gas-cut sulfur water	**O&GC SULW**
oil and gas-cut water	**O&GCW**
oil and salt water	**O&SW**
oil and sulfur water-cut mud	**O&SWCM**
oil and water	**O&W**
oil-based mud	**OBM**
oil circuit breaker	**OCB**
Oil Creek	**Oil Cr**
oil cut	**OC**
oil down to	**ODT**
oil emulsion	**OE**
oil emulsion mud	**OEM**
oil fluorescence	**OFLU**
oil fractured	**oilfract**
oil immersed, water cooled	**OIWC**
oil in hole	**OIH**
oil in place	**OIP**
oil in tanks	**OIT**
oil insulated	**OI**
oil insulated, fan cooled	**OIFC**
oil insulated, self-cooled	**OISC**
oil odor	**OO**
oil pay	**OP**
oil payment interest	**OPI**
Oil Refining Industry Action Committee	**ORIAC**
oil sand	**O sd**
oil show	**OS**

oil stain	**OSTN**
oil standing in drillpipe	**OSIDP**
oil string flange	**OSF**
oil to surface	**OTS**
oil unit	**OU**
oil well flowing	**OWF**
oil well from waterflood	**OWFWF**
oil well gas	**OWG**
oil well shut in	**OSI**
oil-cut mud	**OCM**
oil-cut salt water	**OCSW**
oil-cut water	**OCW**
oil-powered total energy	**OTE**
oil-soluble acid	**OSA**
oil-water contact	**OWC**
old plugback	**OPB**
old plugback depth	**OPBD**
old total depth	**OTD**
old well drilled deeper	**OWDD**
old well plugged back	**OWPB**
old well sidetracked	**OWST**
old well worked over	**OWWO**
olefin	**ole**
Oligocene	**Olig**
on center	**OC**
one thousand foot-pounds	**kip-ft**
one thousand pounds	**kip**
ooliclastic	**ooc**
oolimoldic	**oom**
oolitic	**ool**
open (ed) (ing)	**opn**
open choke	**OC**
open cup	**OC**
open end	**OE**
open flow	**OF**
open flow potential	**OFP**
open hearth	**OH**
open hole	**OH, op hole**
open line (no choke)	**OL**

open tubing	**OT**
open-file report	**OF**
operate, operations, operator	**oper**
operation shutdown	**OSD**
operations and maintenance	**O&M**
operations commenced	**OC**
Operculinoides	**Operc**
opposite	**opp**
optimum bit weight and rotary speed	**OBW & RS**
option to farmout	**optn to F/O**
optional	**OPTL**
orange	**OR**
Ordovician	**Ord**
Oread	**Or**
organic	**org**
organization	**org**
orientation	**ORIENT**
oriented microresistivity	**OMRL**
orifice	**orf**
orifice flange one end	**OFOE**
original oil in place	**OOIP**
original stock tank oil in place	**OSTOIP**
original total depth	**OTD**
original, originally	**orig**
Oriskany	**Orisk**
orthoclase	**orth**
Osage	**Os**
Osborne	**O**
ostracod	**ost**
Oswego	**Osw**
other end beveled	**OEB**
ounce	**oz**
Ouray	**Our**
out of service over and short (report)	**O/S**
out of stock	**O/S**
outboard motor oil	**OBMO**
Outer Continental Shelf	**OCS**
outlet	**otl, OUT**

outline	**OLN**
outpost	**OP**
outside diameter	**OD**
outside screw and yoke (valve)	**OS&Y**
overexpenditure	**OE**
overhead	**OH**
overproduced	**OP**
overall	**OA**
overall height	**OAH**
overall length	**OAL**
overflush (ed)	**OFL**
overhead	**ovhd**
overriding royalty	**ORR**
overriding royalty interest	**ORRI**
overseas procurement office	**OPO**
overshot	**OS**
overtime	**OT**
oxidized, oxidation	**ox**
oxygen	**oxy**

P

Pacific Outer Continental Shelf	**POCS**
packed	**pkd**
packer	**pkr**
packer set at	**PSA**
packing, package (ed)	**pkg(d)**
Paddock	**Padd**
page (before a number)	**p.**
Pahasapa	**Paha**
paid	**pd**
Paint Creek	**PC**
pair	**pr**
paleonotology	**Paleo**
Paleozoic	**Paleo**
Palo Pinto	**Palo P**
Paluxy	**Pal, Pxy**
panel	**pnl**

panel board	**PNL BD**
Panhandle lime	**Pan L**
Paradox	**Para**
Park City	**Park C**
parish	**Ph**
parrafins-olefins-napthenes-aromatics	**PONA**
part, partly	**pt**
partial	**PART**
participating area	**PA**
partings	**prtgs**
partition	**PTN**
partly	**ptly**
parts per billion	**ppb**
parts per million	**ppm**
Pascal	**Pa**
patent (ed)	**pat**
pattern	**patn**
pavement	**pvmnt**
paving	**pav**
Pawhuska	**Paw**
payment	**pymt**
pearly	**prly**
pebble, pebbly	**pbl (y)**
Pecan Gap	**PG**
Pecan Gap chalk	**PGC**
pedestal	**PED**
pelecypod	**plcy**
pelletal, pelletoidal	**pell**
penalty, penalize (ed) (ing)	**penal**
penetration asphalt cement	**Pen A.C.**
penetration index	**PI**
penetration, penetration test	**pen**
Pennsylvanian	**Penn**
Pensky-Martins	**PM**
Pensky-Martins (flash)	**P-M**
per-acre bonus	**PAB**
per-acre rental	**PAR**
percent	**%, pct**
per day	**PD**

per foot	**/ft, pft**
per square inch gauge	**psig, PSIG**
percolation	**perco**
perforate (ed) (ing) (or)	**perf**
perforated casing	**perf csg**
perforating hyper select	**H-SEL**
perforating, Enerjet	**EJ**
perforating, Hyperdome	**HD**
perforating, Hyperdome II	**P-HDII**
perforating, Ultrajet	**ULJ**
performance evaluation and review technique	**PERT**
Performance Number (aviation gas)	**PN**
period	**prd**
peripheral wedge zone	**PWZ**
permanent	**perm**
permanent-type completion	**PTC**
permanently shut down	**PSD**
permeability (vertical direction)	**KV**
permeable (ability)	**perm**
Permian	**Perm**
permit	**prmt**
perpendicular	**perp**
personal and confidential	**P&C**
personnel	**pers**
petrochemical	**petrochem**
petroleum	**pet**
petroleum and natural gas	**P & NG**
petroliferous	**petrf**
Pettet	**Pet**
Pettit	**Pett**
Pettus sand	**Pet. sd**
phase	**ph**
Phosphoria	**Phos**
phrohotite	**po**
picked up	**PU**
picking up drillpipe	**PUDP**
Pictured Cliff	**Pic Cl**
piece	**pc**

pilot	plt
pilot-loaded valve	PLV
pin end	pe
Pin Oak	P.O.
Pine Island	PI
pink	pk
pinpoint	pinpt, PP
pinpoint porosity	PPP
pint	pt
pipe butt weld	PBW
pipe electric weld	PEW
pipe lapweld	PLW
pipe, seamless	PSM
pipe sleeve	PSL
pipe spiral weld	PSW
pipe to soil potential	PTS pot
pipe-handling capacity	PHC
pipeline	PL
pipeline oil	PLO
pipeline terminal	PLT
pipeway	PWY
piping	ppg
piping and instrument diagrams	P&IDS
piping diagram	P/DIA
pisolites, pisolitic	piso
pitted	pit
plagioclase	plg
plain both ends	PBE
plain end	PE
plain end beveled	PEB
plain large end	PLE
plain one end	POE
plain small end	PSE
plan	pln
plan of development	POD
plant	plt
plant (pressure) volume reduction	PVR
plant fossils	pl fos
Planulina harangensis	Plan. hara.
Planulina palmarie	Plan. palm.

plaster	**PLASR**
plastic	**plas**
plastic viscosity	**PV**
plate	**PL**
platform	**platf**
platy	**plty**
please note and return	**PNR**
Pleistocene	**Pleist**
Pliocene	**Plio**
plug	**Pg**
plug down	**PD**
plug on bottom	**POB**
plugged	**plgd**
plugged and abandonded	**P & A**
plugged back	**PB**
plugged-back depth	**PBD**
plugged-back total depth	**PBTD**
plumbing	**PLMB**
plunger	**plngr**
pneumatic	**pneu**
Podbielniak	**Pod.**
point	**pt**
Point Lookout	**Pk Lkt**
poison	**pois**
poker chipped	**PC**
polish (ed)	**pol**
polished rod	**PR**
polyethylene	**poly cl**
polymerization, polymerized	**poly**
polymerized gasoline	**polygas**
polypropylene	**polypl**
polyvinyl chloride	**PVC**
Pontotoc	**Pont**
pooling agreement	**PA**
poor	**pr**
porcelaneous	**porc**
porcion	**porc**
pore volume	**PV**
porosity and permeability	**P&P**
porosity, porous	**por**

porous and permeable	**P&P**
port collar	**pc**
portable	**port**
Porter Creek	**PC**
position	**pos**
positive	**pos**
positive crankcase ventilation	**PCV**
possible (ly)	**poss**
Post Laramie	**P Lar**
Post Oak	**P.O.**
postweld heat treatment	**PWHT**
potable water	**POT/WTR**
potential	**pot**
potential difference	**pot dif**
potential test	**PT**
potential test to follow	**PTTF**
pound	**lb**
pound-inch	**lb-in.**
pounds per barrel	**PPB**
pounds per cubic foot	**PCF**
pounds per foot	**lb/ft**
pounds per gallon	**PPG**
pounds per square foot	**psf, lb/sq ft**
pounds per square inch	**psi, PSI**
pounds per square inch absolute	**psia, PSIA**
pounds per square inch gauge	**psig, PSIG**
pour point (ASTM method)	**pour ASTM**
power	**PWR**
power distribution center	**PDC**
power distribution system	**PDS**
power factor	**PF**
power factor meter	**PFM**
Precambrian	**Pre Camb**
precast	**prcst**
precipitate	**ppt**
precipitation number	**ppn No.**
precipitator	**PRECIP**
predominant	**predom**
prefabricated	**prefab**

preferred	**pfd**
prefractionator	**PFRACT**
preheater	**prehtr**
preliminary	**prelim**
premium	**prem**
prepaid	**ppd**
prepare, preparing, preparation	**prep**
preparing to take potential test	**PRPT**
present depth	**PD**
present operations	**pr op**
present production	**P.P.**
present total depth	**PTD**
present worth at discount rate of 15%	**PW(15)**
pressed distillate	**PD**
pressure	**press**
pressure alarm	**PA**
pressure control valve	**PCV**
pressure differential controller	**PDC**
pressure differential indicator	**PDI**
pressure differential indicator controller	**PDIC**
pressure differential recorder	**PDR**
pressure differential recorder controller	**PDRC**
pressure indicator	**PI**
pressure indicator controller	**PIC**
pressure recorder	**PR**
pressure recorder control	**PRC**
pressure safety valve	**PSV**
pressure seal bonnet	**PSB**
pressure switch	**PS**
pressure-volume-temperature	**PVT**
prestressed	**prest**
prevent, preventive	**prev**
prevention of significant deterioration, EPA	**PSD**
previous	**PREV**
previous daily output average	**PREV DO AVG**

primary	**pri**
primary reference fuel	**PRF**
principal	**prin**
principal lessee (s)	**prncpl lss**
prism (atic)	**pris**
private branch exchange	**PBX**
privilege	**priv**
probable (ly)	**prob**
process	**proc**
process & instrument diagram	**P & ID**
process flow diagram	**PFD**
Process Performance Index	**PPI**
produce (ed) (ing) (tion), product (s)	**prod**
producing gas well	**PGW**
producing oil and gas well	**POGW**
producing oil well	**POW**
producing oil well, flowing	**POWF**
producing oil well, pumping	**POWP**
producing well	**PW**
production department exploratory test	**PDET**
production payment	**PP**
production payment interest	**PPI**
production test flowed	**PTF**
production test pumped	**PTP**
productivity index	**PI**
professional paper	**PP**
profit and loss	**P&L**
profit-sharing interest	**PSI**
progress	**prog**
project (ed) (ion)	**proj**
project ultimate cost	**PUC**
projected total depth	**PTD**
propane	**LPG**
property	**PROP**
property line	**PL**
proportional	**prop**
propose (ed)	**prop**
proposed bottom-hole location	**PBHL**

proposed depth	**PD**
proposed total depth	**PTD**
prorated	**pro**
prospect	**Psp**
protection	**prot**
Proterozoic	**Protero**
provincial	**Prov**
pseudo	**pdso, ps**
public address	**PA**
public relations	**PR**
Public School Land	**PSL**
pull (ed) rods and tubing	**PR&T**
pull (put) out of hole	**POOH**
pulled	**pld**
pulled (put) out of hole	**POH**
pulled bid pipe	**PBP**
pulled out	**PO**
pulled pipe	**PP**
pulled up	**PU**
pulled up in casing	**PUIC**
pulling	**plg**
pulling tubing	**PTG**
pulling tubing and rods	**PTR**
pulsation dampener	**PD**
pulse (sating) (sation)	**PULS**
pump	**P/**
pump and flow	**P&F**
pump building	**P/BLDG**
pump in	**PI**
pump jack	**PJ**
pump job	**PJ**
pump on beam	**POB**
pump pressure	**PP**
pump station	**PS**
pump testing	**P tstg**
pump-in pressure	**PIP**
pump (ed) (ing)	**pmp (d) (g)**
pumper's depth	**PD**
pumping equipment	**PE**
pumping for test	**PFT**

pumping load oil	**PLO**
pumping unit	**PU**
pumps off	**PO**
purchase order	**PO**
purchasing	**PURCH**
purchasing request	**PR**
purification	**PURF**
purple	**purp**
putting on pump	**POP**
pyrite, pyritic	**pyr**
pyrobitumen	**pyrbit**
pyroclastic	**pyrclas**
pyrolysis	**pyls**

—————— **Q** ——————

quadrant	**QDRNT**
quadrant (rangle) (ruple)	**quad**
qualitative	**QUAL**
quality	**qual**
quality assurance	**QA**
quality control	**Q.C., QC**
quality discount allowance	**QDA**
quantity	**qty**
quarry	**qry**
quart (s)	**qt**
quarter	**qtr**
quartz, quartzite, quartzitic	**qtz**
quartzose	**qtzose**
Queen City	**Q. City**
Queen Sand	**Q. sd**
quench	**qnch**
questionable	**quest**
quick ram change	**QRC**
quintuplicate	**quint**

R

rack	**RK**
radiant	**RADT**
radiation	**radtn**
radical	**rad**
radioactive	**RA**
radiographic test	**RT**
radiological	**rad**
radius	**R**
radius	**rad**
railing	**rlg**
railroad	**RR**
Railroad Commission (Texas)	**RRC**
rainbow show of oil	**RBSO**
raised face	**RF**
raised face, flanged end	**RFFE**
raised face, slip on	**RFSO**
raised face, smooth finish	**RFSF**
raised face, weld neck	**RFWN**
Ramsbottom Carbon Residue	**RCR**
ran (running) rods and tubing	**RR&T**
ran in hole	**RIH**
random lengths	**RL**
range	**rge**
Ranger	**Rang**
Rankine (temp. scale)	**R**
rapid curing	**RC**
rat hole	**RH**
rat hole mud	**RHM**
rate of penetration	**ROP**
rate of return	**ROR**
rate too low to measure	**RTLTM**
rating	**rtg**
raw gas	**RG**
raw gas lift	**RAGL**
re-evaluation for overoptimism	**REFOO**
reabsorber	**REABS**
reacidize (ed) (ing)	**reacd**
reaction (ed)	**react**

ready for rig	**RFR**
ream	**rm**
reamed	**rmd**
reaming	**rmg**
reboiler	**RBLR**
received	**recd**
receiver	**recr**
receptacle	**recp,** **RCPT**
reciprocate (ing)	**recip**
recirculate	**recirc**
recommend	**rec**
recommended spare part	**R-SP**
recomplete (ed) (ion)	**recomp**
recompressor	**RECOMP**
recondition (ed)	**recond**
record (er) (ing)	**rec**
recover (ed) (ing), recovery	**rec**
recovery	**RCVY**
rectangle, rectangular	**rect**
rectifier	**rect**
recycle	**recy,** **RCYL**
red beds	**Rd Bds**
Red Cave	**RC**
Red Fork	**Rd Fk**
red indicating lamp	**RIL**
Red Oak	**R.O.**
Red Peak	**Rd Pk**
Red River	**RR**
redrilled	**redrld, RR**
reducer	**RDCR**
reducing balance	**red bal**
reducing, reducer	**red**
reference	**ref**
refine (ed) (er) (ry)	**ref**
Refinery Technology Laboratory	**RTL**
refining	**refg**
reflect (ed) (ing) (tion)	**refl, RFLCT**

reflux	**refl**
reformate (er) (ing)	**reform**
reformer	**REFMR**
refraction, refractory	**refr**
refrigerator (rant) (tion)	**REFRIG**
refrigeration building	**RFG/BD**
regenerator	**regen**
register	**reg, RGTR**
regular acid	**R/A**
regular, regulator	**reg**
Reid vapor pressure	**RVP**
reinforce (ed) (ing) (ment)	**reinf**
reinforced concrete	**reinf conc**
reinforcing bar	**rebar**
reject	**rej**
rejection	**rej'n**
Reklaw	**Rek**
relative humidity	**RH**
relay	**rly**
release (ed) (ing)	**rls (ed) (ing), rel**
released swab unit	**RSU**
relief	**rlf**
relief valve	**RV**
relocate (ed)	**reloc**
remains	**rmns, rems**
remedial	**rem**
remote control	**RC**
remote operating system (station)	**ROS**
remote terminal unit	**RTU**
remove (al) (able)	**rmv (l)**
Renault	**Ren**
rental	**rent**
Reophax bathysiphoni	**Reo. bath.**
repair (ed) (ing) (s)	**rep**
repairman	**rpmn**
reperforated	**reperf**
replace (ed)	**rep**
replace (ment)	**repl**
report	**rep, RPRT**

report available only through National Technical Information Service	**PB-ADA**
	PB-ADA
reported	rptd
request	**REQ**
request for proposal	**RFP**
request for quote	**RFQ**
required	reqd
requirement	reqmt
requisition	req
research	res
research and development	**R & D**
Research Octane Number	**Res. O. N., RON**
Research Planning Institute	**RPI**
reserve (ation)	res
reservoir	rsvr
reservoir description service	**RDS**
residual, residue	resid
resinous	rsns
resistance, resistivity, resistor	res
resistivity	**R**
resistivity as recorded from 16″ electrode configuration	**R(16″)**
resistivity, invaded zone	**RIZ**
resistivity, flushed zone	**Rxo**
resistivity, mud	**Rm**
resistivity, mud filtrate	**Rmf**
resistivity, water	**Rw**
resistivity, water (apparent)	**Rwa**
resistor (s)	**RESIS**
retail pump price	**RPP**
retain (er) (ed) (ing)	ret, rtnr
retard (ed)	rtd
retrievable	retr
retrievable bridge plug	**RBP**
retrievable retainer	retr ret
retrievable test treat squeeze (tool)	**RTTS**
return	ret

return on investment	**ROI**
returnable steel drum	**RSD**
returned	**retd**
returned well to production	**RWTP**
returning circulation oil	**RCO**
reverse (ed)	**rvs, rev (d)**
reverse circulation	**RC**
reverse circulation rig	**RCR**
reversed out	**rev/O, RO**
revise (ed) (ing) (ion)	**rev**
revolution (s)	**rev**
revolutions per minute	**rpm**
revolutions per second	**rps**
rework (ed)	**rwk (d)**
rheostat	**rheo**
ribbon sand	**Rib**
rich oil fractionator	**ROF**
Rierdon rig	**Rier**
rig (ged) (ging) up	**RU**
rig floor	**RF**
rig on location	**ROL**
rig released	**RR, R Rel**
rig repair	**RR**
rig service	**RS**
rig skidded	**RS**
rig time	**RT**
rig-up casing crew	**RUCC**
rigged (ing) down	**RD**
rigged down, moved out	**RDMO**
rigged-down swabbing unit	**RDSU**
rigging rotary	**RR**
rigging-up cable tools	**RUCT**
rigging-up machine	**RUM**
rigging-up pump	**RUP**
rigging-up rotary tools	**RURT**
rigging-up service rig	**RUSR**
rigging-up standard tools	**RUST**
rigging-up swabbing unit	**RUSU**
rigging-up tools	**RUT**
right angle	**RA**

right hand	**RH**
righthand door	**RHD**
right of way	**ROW**
ring	**rg**
ring groove	**RG**
ring joint	**RJ**
ring joint, flanged end	**RJFE**
ring-tool joint	**RTJ**
ring-type joint	**RTJ**
rising stem (valve)	**RS**
rivet	**riv. RVT**
road (s)	**rd (s)**
road & location	**R&L**
road & location complete	**R&LC**
Robulus	**Rob**
rock	**rk**
rock bit	**RB**
rock pressure	**RP**
Rockwell hardness number	**RHN**
rocky	**rky**
Rodessa	**Rod**
rods and tubing	**R & T**
roentgen	**r**
roofing	**RFG**
room	**rm**
root mean square	**RMS**
rose	**ro**
Rosiclare sand	**Ro sd**
rotameter	**RTMTR**
rotary bushing	**RB**
rotary bushing measurement	**RBM**
rotary drive bushing	**RDB**
rotary drive bushing to ground	**RDB-GD**
rotary kelly bushing	**RKB**
rotary table	**RT**
rotary test	**R test**
rotary tools	**RT**
rotary total depth	**RTD**
rotary unit	**RU**
rotary, rotate, rotator	**rot**

rotative gas lift	**ROGL**
rough	**rgh**
rough order of magnitude	**ROM**
round	**rd**
round thread	**rd thd**
round trip	**rdtp**
round trip changed bit	**RT CB**
rounded	**rdd, rnd**
routing	**RTG**
rows	**R**
royalty	**roy**
royalty interest	**RI**
rubber	**rbr, rub**
rubber ball sand water frac	**RBSWF**
rubber ball sand oil frac	**RBSOF**
rubber balls	**Rbls**
run of mine	**ROM**
running	**rng**
running casing	**RC**
running electric log	**REL**
running radioactive log	**RALOG**
running tubing	**RTG**
rupture	**rupt**
rust and oxidation	**R&O**

S

Sabinetown	**Sab**
saccharoidal	**sach**
sack (s)	**sk, sx**
saddle	**sadl**
Saddle Creek	**Sad Cr**
safety	**saf**
safety relief valve	**SRV**
safety/department	**SAF/DPT**
Saint Genevieve	**St Gen**
Saint Louis lime	**St L**
Saint Peter	**St Ptr**

Salado	**Sal**
salary, salaried	**sal**
Saline Bayou	**Sal Bay**
salinity	**sal**
salt	**X**
salt and pepper	**s&p**
Salt Mountain	**Slt Mtn.**
salt wash	**SW**
salt water	**SW, swtr, XW**
salt water to surface	**SWTS**
saltwater-cut mud	**SWCM**
saltwater disposal	**SWD**
saltwater disposal system	**SWDS**
saltwater disposal well	**SWDW**
saltwater fracture	**SWF**
saltwater injection	**SWF**
salty	**Slty**
salty sulfur water	**SSUW**
salvage	**salv**
sample	**samp, smpl, spl**
sample chamber	**spl cham**
sample formation tester	**SFT**
sample tops	**S/T**
San Andres	**San And**
San Angelo	**San Ang**
San Bernardino base and meridian	**SBB&M**
San Rafael	**San Raf**
Sanastee	**Sana**
sand	**SD, sd**
sand and shale	**sd & sh**
sand-oil fracked	**sdoilfract**
sand-oil fracture	**SOF**
sand showing gas	**Sd SG**
sand showing oil	**Sd SO**
sand-water fracked	**sdwtrfract**
sanded	**sdd**
sandfracked	**sdfract, SF**

sandstone	**SS**
sandy	**sdy**
sandy lime	**sdy li**
sandy shale	**sdy sh**
sanitary	**SAN, sani**
sanitary water	**S/WTR**
Santa Margarita	**Sta Marg**
saponification	**sap**
saponification number	**Sap No.**
Saratoga	**Sara**
Satanka	**Stṇka**
saturated, saturation	**sat**
Sawatch	**Saw**
Sawtooth	**Sawth**
Saybolt furol	**Say furol**
Saybolt Seconds Universal	**SSU**
Saybolt universal viscosity	**SUV**
scaffolding	**SCAF**
scales	**sc**
scatter (ed)	**scatt (d), sctrd**
schedule	**sch**
schematic	**schem**
Schlumberger	**SCHL**
scolescodonts	**scolc**
scraper	**scr**
scratcher	**scr**
screen	**scr**
screw (ed)	**scr (d)**
screw end American National Acme thread	**SE NA**
screw end American National Coarse thread	**SE NC**
screw end American National Taper Pipe thread	**SE NTP**
screwed and socketweld	**S/SW**
screwed end	**S/E**
screwed on one end	**SOE**
scrubber	**scrub**
sea level	**S.L.**

Seabreeze	**Sea**
seal assembly	**SA**
seal oil	**SEO**
seal weld	**SWLD**
seal-welded bonnet	**SWB**
sealed	**sld**
seamless	**smls**
seating nipple	**SN**
secant	**sec**
second (ary)	**sec**
secondary butyl alcohol	**SBA**
seconds	**S, sec**
secretary	**sec**
section	**sec**
section (s) (al) (ing)	**SECT**
section line	**SL**
section-township-range	**S-T-R**
securaloy	**scly**
sediment (s)	**sed**
Sedwick	**Sedw**
segment	**SEG**
seismograph, seismic	**seis**
selection (tive) (tor)	**SELECT**
selenite	**sel**
self (spontaneous) potential	**SP**
self-contained	**SC**
self-elevating work platform	**SEWOP**
Selma	**Sel**
Senora	**Sen**
separate, separator, separation	**SEP**
septuplicate	**sept**
sequence	**seq**
series, serial	**ser**
serpentine	**serp**
Serratt	**Serr**
service (s)	**svc, serv**
service charge	**serv chg**
service unit	**svcu**
set drillpipe	**SDP**
set plug	**SP**
settling	**set**

Seven Rivers	**S Riv**
service station	**SS**
severy	**Svry**
Seward Meridian (Alaska)	**SM**
sewer	**sew**
sexton	**Sex**
sextuple	**sxtu**
sextuplicate, sextuplet	**sext**
shaft horsepower	**shp**
shake out	**SO**
shale	**sh**
shaled out	**SO**
shaly	**shly**
shallower pool (pay) test	**SPT**
shallower pool wildcat, discovery	**WSD**
shallower pool wildcat, dry	**WS**
Shannon	**Shan**
shear	**shr**
sheathing	**shthg**
sheet	**sh**
sheeting	**SHTG**
shell and tube	**S & T**
shells	**shls**
shelter	**SHLT**
Shinarump	**Shin**
ship (ping)	**shp(g)**
shipping point (purchasing term)	**s/p, sp**
shipment	**shpt**
shock sub	**SS**
shop fabrication	**S/FAB**
short radius	**SR**
short string	**SS**
short thread	**ST**
short threads & coupling	**ST&C**
shortage	**SHTG**
shot open hole	**SOH**
shot per foot	**SPF**
shot point	**sp**
shoulder	**shld**

show condensate	**SC**
show gas	**SG**
show gas and condensate	**SG&C**
show gas and distillate	**SG&D**
show gas and water	**SG & W**
show of dead oil	**SDO**
show of free oil	**SFO**
show of gas and oil	**SG&O**
show of oil	**SO**
show oil and gas	**SO&G**
show oil and water	**SO&W**
shut down	**SD**
shut down awaiting orders	**SDWO**
shut down for orders	**SDO**
shut down for pipe line	**SDPL**
shut down for repairs	**SDR**
shut down for weather	**SDW**
shut down overnight	**SDON**
shut down to acidize	**SDA**
shut down to fracture	**SDF**
shut down to log	**SDL**
shut down to plug & abandon	**SDPA**
shut in	**SI**
shutin bottom-hole pressure	**SIBHP**
shutin casing pressure	**SICP**
shutin gas well	**SIGW**
shutin oil well	**SIOW**
shutin pressure	**SIP**
shutin tubing pressure	**SITP**
shutin wellhead pressure	**SIWHP**
shutin, waiting on potential	**SIWOP**
shut well in overnight	**SWION**
side door choke	**SD Ck**
side opening	**SO**
sidewall cores	**SWC**
sidewall neutron porosity	**SWNP**
sideboom	**SB**
siderite (ic)	**sid**
sides, tops & bottoms	**s,t&b**
sidetrack (ed) (ing)	**sdtkr, ST**

sidetracked hole	**STH**
sidetracked total depth	**STTD**
sidewall	**SDWL**
sidewall samples	**SWS**
siding	**SDG**
signed	**sgd**
silencer	**SLNCR**
silica, siliceous	**silic**
silky	**slky**
siltstone	**silt**
silty	**slty**
Silurian	**Sil**
similar	**sim**
Simpson	**Simp**
single (s)	**sgl (s)**
single-pole double throw	**SPDT**
single-pole double throw switch	**SPDT SW**
single-pole single throw	**SPST**
single-pole single throw stitch	**SPST SW**
single-pole switch	**SP SW**
single random lengths	**SRL**
single shot	**SS**
Siphonina davisi	**Siph. d.**
size	**sz**
sketch	**SK**
skimmer	**skim**
Skinner	**Skn**
Skull Creek	**Sk Crk**
sleeper	**SLPR**
sleeve	**sl, SLV**
sleeve bearing	**SB**
slickensided	**sks**
sliding-scale royalty	**S/SR**
slight (ly)	**sli**
slight oil-cut mud	**SOCM**
slight oil-cut water	**SOCW**
slight show of gas	**SSG**
slight show of oil	**sli SO, SSP**
slight show of oil and gas	**SSO&G**
slight, weak, poor fluorescence	**SFLU**

slightly gas-cut salt water	**SGCSW**
slightly gas-cut water cushion	**SGCWC**
slightly gas-cut mud	**SGCM**
slightly gas-cut oil	**SGCO**
slightly gas-cut water	**SGCW**
slightly gas-cut water blanket	**SGCWB**
slightly oil- and gas-cut mud	**SO&GCM**
slightly oil-cut salt water	**SOCSW**
slightly oil-cut water blanket	**SOCWB**
slightly oil-cut water cushion	**SOCWC**
slightly porous	**sp**
Sligo	**Sli**
slim-hole drillpipe	**SHDP**
slip and cut drill line	**SC DL**
slip on	**SO**
slope type of wall to keep out flooding	**berm**
slow set (cement)	**SS**
slurry	**slur**
Smackover	**Smk, SO**
small	**sm**
small show	**SS**
Smithwick	**Smithw**
Smoke Volatility Index	**SVI**
smooth	**smth**
snubber, snubbing	**SNUB**
snuffing	**SNUFF**
Society of Economic Paleontologists & Mineralogists	**SEPM**
Society of Petroleum Engineers	**SPE**
socket	**skt**
socket weld	**SW**
sodium-base grease	**sod gr**
sodium carbonate	**NaCO$_3$**
sodium carboxymethylcellulose	**CMC**
sodium chloride	**NaCL**
sodium hydroxide	**NaOH**
soft	**sft**
solar heat medium	**SHM**
solenoid	**sol, slnd**

solenoid-operated valves	**SOV**
solenoid valve	**SV**
solids	**sol**
solution	**soln**
solvent	**solv**
somastic	**som**
somastic coated	**somct**
sonic log	**SONL**
sort (ed) (ing)	**srt**
south	**S**
south half	**S/2**
south line	**SL**
south offset	**S O**
southeast	**SE**
southeast corner	**SE/C**
southeast quarter	**SE/4**
southerly	**S'ly**
southwest	**SW**
southwest corner	**SW/c**
southwest quarter	**SW/4**
spacer	**spcr**
spare	**sp**
Sparta	**Sp**
spearfish	**spf**
special	**spcl**
specialty	**splty**
specific gravity	**sp gr**
specific heat	**sp ht**
specific volume	**sp. vol.**
specification	**spec**
speckled	**speck**
speed/current	**S/C**
speed/torque	**S/T**
Sphaerodina	**Sphaer**
sphalerite	**sphal**
spherules	**sph**
spicule (ar)	**spic**
spigot	**SPGT**
spigot and spigot	**s & s**
spillway	**SPWY**

spindle	spdl
Spindletop	Spletp
spiral weld	SW
spirifers	sprf
Spiroplectammina barrowi	Spiro. b.
splintery	splty
splitter	SPLTR
sponge	spg
spore	sp
spot sales agreement	SSA
spotted	sptd
spotty	sptty
Spraberry	Spra
spring	spg
Springer	Sprin
sprinkler	spkr
sprocket	spkt
spud (ded) (der)	spd
square	sq
square centimeter	sq cm
square foot (feet)	ft^2, sq ft
square inch	sq in.
square kilometer	sq km
square meter	sq m
square millimeter	sq mm
square root	SQRT
square yard (s)	sq yd
squeeze (ed) (ing)	sqz
squeeze packer	sq pkr
squeezed	sq
squirrel cage	sq cg
stabilized (er)	stab
stage	STG
staggered	STAG
stain (ed) (ing)	stn (d) (g)
stain and odor	S&O
stainless steel	SS
stairway	stwy
Stalnaker	Stal
stand (s) (ing)	std

stand by	**stn/by**
standard cubic feet per day	**SCFD, scfd**
standard cubic feet per hour	**SCFH, scfh**
standard cubic feet per minute	**SCFM, scfm**
standard cubic foot	**SCF, scf**
standard operational procedure	**SOP**
standard temperature and pressure	**STP**
standardization	**STDZN**
standards	**std (s) (g)**
standing	**stdg**
Stanley	**Stan**
start	**st**
start of cycle	**SOC**
start of run	**SOR**
started in hole	**SIH**
started out of hole	**SOH, SOOH**
starting fluid level	**SFL**
state lease	**SL**
state potential	**State pot**
statewide rules	**SWR**
static bottom-hole pressure	**SBHP**
station	**sta**
stationary	**stat**
statistical	**stat**
steady	**stdy**
steam	**stm**
steam cylinder oil	**stm cyl oil**
steam emulsion number	**SE No.**
steam engine oil	**stm eng oil**
steam trace (ing)	**STM TR**
steam working pressure	**SWP**
steel	**stl**
steel line correction	**SLC**
steel line measurement	**SLM**
steel tape measurement	**STM**
Steele	**Stel**

stencil (ed) (ing)	**stncl (d) (g)**
stenographer	**steno**
Stensvad	**Stens**
sticky	**stcky**
stiffener	**STIF**
stippled	**stip**
stirrup	**stir**
stock	**stk**
stock tank barrels	**STB, stb**
stock tank barrels per day	**STBPD, stb/d**
stock tank oil in place	**stoip**
stock tank vapor	**STV**
Stone Corral	**Stn Crl**
Stony Mountain	**Sty Mtn**
stopper (ed)	**stpr (d)**
storage	**strg**
stove oil	**stv**
straddle	**strd**
straddle packer	**SP**
straddle packer drillstem test	**SP-DST**
straight	**strt**
straight-hole test	**SHT**
straight-run naphtha	**SRN**
straightened	**strtd**
straightening	**stging**
strainer	**stnr**
strand (ed)	**strd**
strap out of hole	**STROH**
strapped out of hole	**SOOH**
stratigraphic	**strat**
Strawn	**Str**
streak (s) (ed)	**stk, strk**
striated	**stri**
string (er)	**strg (r)**
string shot	**SS**
strip (per) (ping)	**STPR**
strokes per minute	**SPM**
stromatoporoid	**strom**
strong	**strg**

structure, structural	**struc**
stuck	**stk**
study	**stdy**
stuffing box	**SB**
styolite, styolitic	**styo**
sub-Clarksville	**Sub Clarks**
subangular	**sub angl**
subdivision	**subd**
subrounded	**sub rnd**
subsea	**SS**
subsidiary	**sub**
substance	**sub**
substation	**substa**
substitute	**SUBST**
substructure height	**SH**
subsurface	**SS**
subsurface safety valve	**SSSV**
successful wildcat outpost	**WOE**
sucker rod	**skr rd**
sucrose, sucrosic	**suc**
suction	**suct**
sugary	**sug**
sulfate bacteria	**SRB**
sulfur by bomb method	**S Bomb**
sulfur, sulfuric	**sulf**
sulfuric acid	**H_2SO_4**
sulfated	**sulf**
sulfur	**sul**
sulfur water	**sul wtr**
summary, summarize	**sum**
Summerville	**Sumvl**
Sunburst	**Sb, Sunb**
Sundance	**Sund**
Supai	**Sup**
superintendent	**supt**
superseded	**supsd**
supervisor	**suprv**
supplement	**supp**
supply (ied) (ier) (ing)	**supl, sply**
support	**sppt**

surface	**surf, sfc**
surface approximation and formation evaluation	**SAFE**
surface-controlled subsurface safety valve	**SCSSV**
surface flow pressure	**SFP**
surface geology	**SG**
surface measurement	**SM**
surface pressure	**SP**
surge	**SRG**
surplus	**surp**
survey	**sur**
suspended	**susp**
suspended ceiling	**SUSP CLG**
swab and flow	**S&F**
swab rate	**SR**
swab run (s)	**SR**
swabbed	**S/**
swabbed, swabbing	**SWB**
swabbing unit	**SWU**
swaged	**swd**
Swastika	**Swas**
sweetening	**Swet**
switch	**SW**
switchboard	**PBX, swbd**
switchgear	**swgr**
switchrack	**SWRK**
Sycamore	**Syc**
Sylvan	**Syl**
symbol	**sym**
symmetrical	**sym**
synchronous, synchronizing	**syn**
synchronous converter	**syn conv**
synchroscope	**SYNSCP**
synthesis	**SYNTH**
synthetic	**syn**
synthetic natural gas	**SNG**
system	**sys**
system flow diagram	**SFD**

T

tabular, tabulating	**tab**
tachometer	**TACH**
tag closed cup (flash)	**TCC**
tag open cup (flash)	**TOC**
Tagliabue	**Tag**
Tallahatta	**Tal**
Tampico	**Tamp**
tangent	**tan**
tank	**tk**
tank battery	**TKB**
tank car	**T/C**
tank farm	**TKF**
tank truck	**TT**
tank wagon	**TKW**
tankage	**tkg**
tanker (s)	**tkr**
Tannehill	**Tann**
Tansill	**Tan**
taper pipe thread	**TPT**
Tar Springs sand	**TSS**
Tarkio	**Tark**
tarred and wrapped	**T&W**
taste	**tste**
Taylor	**Tay**
tearing out rotary tools	**TORT**
technical, technician	**tech**
techniques of water-resources investigations	**TWI**
tee	**T**
teeth	**T**
telegraph	**TLG**
Telegraph Creek	**Tel Cr**
teletype	**TWX**
television	**TV**
telephone, telegraph	**tel**
temperature	**Temp**
temperature control valve	**TCV**

temperature controller	**TC**
temperature differential indicator	**TDI**
temperature differential recorder	**TDR**
temperature gradient	**TG**
temperature indicator	**TI**
temperature indicator controller	**TIC**
temperature observation	**TO**
temperature recorder	**TR**
temperature recorder controller	**TRC**
temperature survey indicated top cement at	**TSITC**
temporarily abandoned	**TA**
temporarily shut down	**TSD**
temporarily shut in	**TSI**
temporary (ily)	**temp**
temporary dealer allowance	**TDA**
temporary voluntary allowance	**TVA**
tender	**tndr**
tensile strength	**tens str, TS**
Tensleep	**Tens**
tentaculites	**tent**
tentative	**tent**
Teremplealeau	**Tremp**
terminal board	**T/BRD**
terminal box	**T/Box**
terminate (ed) (ing) (ion)	**termin**
tertiary	**Ter**
tertiary butyl alcohol	**TBA**
test (er) (ing)	**tst (r) (g)**
test to follow	**TTF**
testing on pump	**TOP**
tetraethyl lead	**TEL**
tetramethyl lead	**TML**
Texana	**Tex**
Textularia articulata	**Text. art.**
Textularia dibollensis	**Text. d.**
Textularia hockleyensis	**Text. h.**
Textularia warreni	**Text. w.**
texture	**tex**

Thaynes	**Thay**
thence	**th**
theoretical production and allocation	**TP&A**
thermal	**thrm**
thermal cracker	**therm ckr**
thermal decay time	**TDT**
thermal hydrodealkylation	**THD**
thermofor catalytic cracking	**TCC**
thermometer	**therm**
Thermopolis	**Ther**
thermostat	**therst**
thick, thickness	**thk**
thin bedded	**TB**
thousand (i.e., 13K = 13,000)	**K**
thousand (i.e., 9M = 9,000)	**M**
thousand barrels fluid per day	**MBF/D, MBFPD**
thousand barrels of oil per day	**MBO/D, MBOPD**
thousand barrels of water per day	**MBW/D, MBWPD**
thousand British thermal units	**MBtu**
thousand cubic feet of gas per day	**MCFGPD, Mcfgpd**
thousand cubic feet per day	**MCFD, Mcfd**
thousand electron-volts	**kev**
thousand gallons	**MG**
thousand standard cubic feet	**MSCF, Mscf**
thousand standard cubic feet per day	**MSCF/D, Mscf/d**
thousand standard cubic feet per hour	**MSCF/H, Mscf/h**
thread large end	**TLE**
thread on both ends	**TOBE**
thread small end	**TSE**
thread small end, weld large end	**TSE-WLE**
thread, threaded	**thd**

threaded and coupled	**T & C**
threaded both ends	**TBE**
threaded one end	**TOE**
threaded pipe flange	**TPF**
Three Finger	**Tfing**
Three Forks	**Tfks**
three-phase	**3 PH**
throttling	**thrling**
through	**thru**
through-tubing	**TT**
through-tubing caliper	**TTC**
through-tubing plug	**TTP**
Thurman	**Thur**
tight	**ti, tite**
tight hole	**TH**
tight no show	**TNS**
time delay	**TD**
Timpas	**TIM**
Timpoweap	**Timpo**
tires, batteries, and accessories	**TBA**
to be conditioned for gas	**TRG**
to be conditioned for oil	**TRO**
Todilto	**Tod**
tolerance	**tol**
toluene	**tolu**
ton (after number - 3T)	**T**
tongue and groove (joint)	**T&G**
Tonkawa	**Tonk**
tons	**tons**
too small to measure	**TSTM**
too wet (weak) to measure	**TWTM**
tool (s)	**tl**
tool closed	**TC**
tool joint	**tl jt**
tool open	**TO**
toolpusher	**TP**
tooth	**T**
top and bottom	**T & B**
top and bottom chokes	**T&BC**
top choke	**TC**

top hole choke	**THC**
top hole flow pressure	**THFP**
top of (a formation)	**T/**
top of cement	**TOC**
top of cement plug	**TOCP**
top of fish	**TOF**
top of liner	**TOL**
top of liner hanger	**TLH**
top of pay	**T/pay**
top of salt	**TOS**
top of sand	**T/sd**
top salt	**T/S**
Topeka	**Tpka**
topo sheet evaluation	**TS**
topographic, topography	**topo**
topping	**tpg**
topping and coking	**T & C**
Toronto	**Tor**
Toroweap	**Toro**
torque	**TRQ**
total	**tot**
total depth	**TD**
total time lost	**TTL**
totally enclosed, fan cooled	**TEFC**
tough	**gh**
Towanda	**Tow**
tower	**TWR**
township	**twp**
township (as T2N)	**T**
townsite	**twst**
trace	**TR**
trackage	**trkg**
tract	**TR**
Trans-Alaska Pipeline System	**TAPS**
transducer	**TRNDC**
transfer (ed) (ing)	**trans**
transformer	**trans,**
translucent	**transl**
transmission	**trans**

transmitter	**XMTR**
transparent	**transp**
transportation	**transp**
travel (ed) (ing)	**TRVL**
Travis Peak	**TP**
treat (er) (ed) (ing)	trt (r) (d) (g)
treater	trtr
treating pressure	**TP**
Trenton	**Tren**
Triassic	**Tri**
tricresyl phosphate	**TCP**
trillion	10^{12}
trillion cubic feet	**TCF, Tcf**
trillion cubic feet per day	**TCF/D, Tcf/d**
trilobite	trilo
Trinidad	**Trin**
trip for new bit	**TFNB**
trip in hole	**TIH**
trip out of hole	**TOH**
trip (ped) for bit	**TFB**
triple-pole single throw switch	**3P ST SW**
triple-pole switch	**3P SW**
triplicate	trip
Tripoli	**Trip**
tripolitic	trip
tripped (ing)	trip
truck	trk
true boiling point	**TBP**
true vapor process	**TVP**
true vertical depth	**TVD**
tube	tb
tube bundle	**TB/BDL**
tubing	tbg
tubing and casing cutter	**TCC**
tubing and rods	**T&R**
tubing choke	tbg chk, TC
tubing pressure	**TP, tbg press**

tubing pressure, shut in	**TPSI**
tubing pressure, closed	**TPC**
tubing pressure, flowing	**TPF**
tubinghead flange	**THF**
Tucker	**Tuck**
tuffaceous	**tfs, tuf**
Tulip Creek	**Tul Cr**
tungsten carbide	**tung carb**
turbine compressor	**T/C**
turbo, turbine	**TURB**
turn around	**TA**
turned over to producing section	**TOPS**
turned to test tank	**TTTT**
turnpike	**tpk**
Tuscaloosa	**Tus**
Twin Creek	**Tw Cr**
twisted off	**twst off**
type	**ty**
typewriter	**tywr**
typical	**typ**

U

ultimate	**ult**
ultrahigh frequency	**UHF**
ultrasonic examination	**UT**
ultrasonic test	**UST**
ultraviolet	**UV**
umbrella (s)	**UMB**
Umiat Meridian (Alaska)	**UM**
unbalanced	**UNBAL**
unbranded	**unbr**
unclassified	**U**
unconformity	**unconf**
unconsolidated	**uncons**
under construction	**U/C**
under digging	**UD**
under gauge	**UG**

underreaming	**UR**
underground	**UG**
undifferentiated	**undiff**
unfinished	**unf**
unified coarse thread	**UNC**
unified fine thread	**UNF**
uniform	**uni**
uninterruptible power supply	**UPS**
Union Oil Company	**UOCO**
Union Valley	**UV**
unit	**un**
United States gauge	**USG**
universal gear lubricant	**UGL**
universal transverse mercator	**UTM**
university, universal	**Univ**
unloader	**UNLDR**
unloading	**UNLD**
unsulfonated residue	**UR**
upper (i.e., U/Simpson)	**U/**
upper and lower	**U/L**
upper casing	**UC**
upper tubing	**UT**
upthrown	**UT**
use customer's hose	**UCH**
used rod	**UR**
used with	**U/W**
utility	**UTL**
utility flow diagram	**UFD**
utility water	**U/WTR**
Uvigerina lirettensis	**Uvig. lir.**

---------------- **V** ----------------

vacant	**vac**
vacation	**vac**
vacuum	**vac**
Vaginulina regina	**Vag. reg**
Valera	**Val**
valve	**V, vlv**

Vanguard	**Vang**
vapor pressure	**VP**
vapor recovery	**VR**
vapor recovery unit	**VPU**
vapor temperature	**vt**
vapor (izor)	**vap (r)**
vapor-liquid ratio	**V/L**
varas	**vrs**
variable, various	**var**
variegated	**vari**
varnish	**VARN**
varnish makers and painters naphtha	**VM&P naphtha**
varved	**vrvd**
velocity	**vel**
velocity survey	**V/S, VS**
vendor drawing	**V/DWG**
ventilator	**vent**
Verdigris	**Verd**
Vermillion Cliff	**Ver Cl**
versus	**vs**
vertebrate	**vrtb**
vertical	**vert, vrtl**
vertical support member	**VSM**
very (as very tight)	**v.**
very common	**v.c.**
very fine-grain (ed)	**vfg**
very heavily (highly) gas-cut mud	**VHGCM**
very heavily (highly) gas-cut salt water	**VHGCSW**
very heavily (highly) gas-cut water	**VHGCW**
very heavily (highly) oil- and gas-cut salt water	**VHO&GCSW**
very heavily (highly) oil- and gas-cut mud	**VHO&GCM**
very heavily (highly) oil- and gas-cut water	**VHO&GCW**
very heavily (highly) oil-cut mud	**VHOCM**

very heavily (highly) oil-cut salt water	**VHOCSW**
very heavily (highly) oil-cut water	**VHOCW**
very heavily oil-cut mud	**VHOCM**
very high frequency	**VHF**
very light amber cut	**VLAC**
very noticeable	**v.n.**
very poor sample	**V.P.S.**
very rare	**v.r.**
very slight	**v-sli**
very slight show of gas	**VSSG**
very slight show of oil	**VSSO**
very slightly gas-cut salt water	**VSGCSW**
very slightly gas-cut water	**VSGCW**
very slightly gas-cut mud	**VSGCM**
very slightly oil- and gas-cut mud	**VSO&GCM**
very slightly oil- and gas-cut salt water	**VSO&GCSW**
very slightly oil-cut mud	**VSOCM**
very slightly oil-cut salt water	**VSOCSW**
very slightly oil-cut water	**VSOCW**
very slightly porous	**VSP**
vesicular	**ves**
vessel	**VESS**
vibrate (tor) (ing)	**VIB**
Vicksburg	**Vks**
Viola	**Vi**
Virgelle	**Virg**
viscosity	**vis, V**
viscosity index	**VI**
viscosity-gravity constant	**VGC**
visible	**vis**
vitreous	**vit**
vitrified clay pipe	**VCP**
Vogtsberger	**Vogts**
volatile organic compounds	**VOC**
volt	**V, v**
volt-ampere	**va**
volt-ampere reactive	**var**

voltage	**VOLT**
volume	**V, vol**
volume-percent	**v%**
volumetric efficiency	**vol. eff.**
volumetric subcommittee	**VSC**
vuggy	**vug**
vugular	**vug**

W

Wabaunsee	**Wab**
Waddell	**Wad**
waiting	**wtg**
waiting on	**WO**
waiting on acid	**WOA**
waiting on allowable	**WOA**
waiting on battery	**WOB**
waiting on cable tools or completion tools	**WOCT**
waiting on completion rig	**WOCR**
waiting on drillpipe	**WODP**
waiting on geologist	**WOG**
waiting on orders	**WOO**
waiting on permit	**WOP**
waiting on pipe	**WOP**
waiting on pipeline	**WOPL**
waiting on plastic	**WOP**
waiting on potential test	**WOPT**
waiting on production equipment	**WOPE**
waiting on pump	**WOP**
waiting on pumping unit	**WOPU**
waiting on rig or rotary	**WOR**
waiting on rotary tools	**WORT**
waiting on standard tools	**WOST**
waiting on state potential	**WOSP**
waiting on tank and connection	**WOT&C**
waiting on test or tools	**WOT**
waiting on weather	**WOW**
wall (if used with pipe)	**W**

Wall Creek	**W Cr**
wall thickness (pipe)	**WT**
Waltersburg sand	**Wa sd**
Wapanucka	**Wap**
warehouse	**whse**
Warsaw	**War**
Wasatch	**Was**
wash (ing)	**wsh (g)**
wash and ream	**W&R**
wash and ream to bottom	**WRTBw**
wash oil	**WO**
wash over	**WO**
wash pipe	**WP**
wash to bottom	**WTB**
wash water	**WW**
washed	**w shd**
washing in	**WI**
Washita	**Wash**
Washita-Fredericksburg	**W-F**
washout	**wo**
washover string	**WOS**
waste	**WSTw**
water blanket	**WB**
water closet	**WC**
water cooler	**W/CLR**
water cushion (DST)	**wtr. cush, WC**
water cushion to surface	**WCTS**
water cut	**WC**
water depth	**WD**
water disposal well	**WD**
water in hole	**WIH**
water injection	**WI**
water injection well	**WIW**
water load	**W/L**
water loss	**WL**
water not shut off	**WNSO**
water oil or gas	**WOG**
water saturation	**WS**
water separation index modified	**WSIM**

water shutoff no good	**WSONG**
water shutoff OK	**WSOOK**
water shutoff	**WSO**
water source wells	**WST**
water supply well	**WSW**
water to surface	**WTS**
water well	**WW**
water with slight show of oil	**W/SSO**
water with sulfur odor	**W/sulf O**
water-alternating gas (or water and gas)	**WAG**
water-cut mud	**WCM**
water-cut oil	**WCO**
water-oil ratio	**WOR**
water-supply paper	**WSP**
water-up to	**WUT**
water, watery	**wtr (y)**
waterflood	**WF**
waterproof	**WTR/PRF**
watertight	**WTR/T**
wating on cement	**WOC**
watt	**w**
watt-hour	**w-hr**
weak	**wk**
weak air blow	**WAB**
weather	**wthr**
weather (ed)	**wthd**
weatherproof	**WTHR/PRF**
Weber	**Web**
week	**wk**
weight	**wgt., wt**
weight averaged catalyst temperature	**WACT**
weight on bit	**W.O.B.**
weight-percent	**wt%**
weld ends	**WE**
weld neck	**WN**
welded, welding	**wld**
welder	**wldr**
welding detail (s)	**WLD/DET**

Welex	**Wx**
wellhead injection pressure	**WHIP**
well lines	**WL**
well pad	**WP**
well pad manifolding	**WPM**
wellbore	**wlbr**
wellhead	**WH**
Wellington	**Well**
went back in hole	**WBIH**
west	**W**
west half	**W/2**
west line	**WL**
west offset	**W/O**
westerly	**W'ly**
wet bulb	**WB**
wheel	**WHL**
whipstock	**whip, WS**
whipstock depth	**WSD**
white	**wht**
white dolomite	**Wh Dol**
White River	**WR**
White Sand	**Wh Sd**
wholesale	**whsle**
Wichita	**Wich.**
Wichita Albany	**Wich Alb**
wide	**W**
wide flange	**WF**
Wilcox	**Wx**
wildcat	**WC**
wildcat field, discovery	**WFD**
wildcat outpost, dry	**WO**
Willberne	**Willb**
Wind River	**Wd R**
Windfall Profit Tax	**WPT**
Winfield	**Winf**
Wingate	**Wing**
Winnipeg	**Winn**
Winona	**Win**
wireline	**WL**
wireline coring	**WLC**

wireline test	**WLT**
wireline total depth	**WLTD**
wiring	**WRG**
wiring diagram	**WD**
with	**w/**
without	**W/O**
Wolfcamp	**Wolfc**
Wolfe City	**WC**
Woodbine	**WB**
Woodford	**Wdfd, Woodf**
Woodside	**Woodsd**
work in place	**WIP**
work order	**WO**
worked	**wkd**
working	**wkg**
working interest	**WI**
working pressure	**WP**
workover	**WO, wko**
workover rig	**wkor**
worldscale	**WS**
wrapper	**wpr**
Wreford	**Wref**
wrought iron	**WI**

——————— **X** ———————

X-ray	**X-R**

——————— **Y** ———————

yard (s)	**yd**
Yates	**Y**
Yazoo	**Yz**
year	**yr**

yellow	**yel**
yellow indicating lamp	**YIL**
yield point	**YP**
Yoakum	**Yoak**
your message of date	**YMD**
your message yesterday	**YMY**

Z

zenith	**zen**
Zilpha	**Zil**
zinc	**ZN**
zone	**Zn**

MISCELLANEOUS

eight round pipe	**8rd**
four-pole single throw switch	**4P ST SW**
four-pole switch	**4P SW**
three phase	**3 PH**
trillion	**10^{12}**
triple-pole single throw switch	**3P ST SW**
triple-pole switch	**3P SW**
12 gauge wire-wrapped screen (in a liner)	**12GA W.W.S.**

ABBREVIATIONS FOR LOGGING
TOOLS AND SERVICES

The appropriate companies and associations have not yet established standard abbreviations for the logging segment of the oil and gas industry. The following lists by individual companies are supplemented by a Miscellaneous Section, for your convience.

DRESSER-ATLAS

Acoustilog	ALC
Acoustilog Caliper Gamma Ray	ALC-GR
Acoustilog Caliper Neutron	ALC-N
Acoustilog Caliper Gamma Ray-Neutron	ALC-GRN
Acoustic Cement Bond	CBL
Acoustic Cement Bond Gamma Ray	CBL-GR
Acoustic Cement Bond Neutron	CBL N
Acoustic Cement Bond G/R Neutron	CBL GRN
Acoustic Parameter-Depth	AC PAR D
Acoustic Parameter-Logging	AC PAR L
Acoustic Parameter-16mm Scope	AC PAR 16
Acoustic Signature	AC SIGN
Atlantic Chlorinlog	A CHL
Borehole Compensated	BHC
BHC Acoustilog Caliper	BHC ALC
BHC Acoustilog Caliper Gamma Ray	BHC ALC GR
BHC Acoustilog Caliper Neutron	BHC ALC N
BHC Acoustilog Caliper G/R Neutron	BHC ALC GRN
BHC Acoustilog Caliper (Thru Casing)	BHC AL TC
BHC Acoustilog Caliper Gamma Ray (Thru Casing)	BHC AL GR TC
BHC Acoustilog Caliper G/R Neutron (Thru Casing)	BHC AL GRN TC
Caliper	CL
Casing Potential Profile	CPP
Cemotop	CTL
Channelmaster	CML
Channelmaster-Neutron	CML N
Chlorinlog	CHL
Chlorinlog-Gamma Ray	CHL GR
Compensated Densilog Caliper	C DLC

Compensated Densilog Caliper Gamma Ray	C DLC GR
Compensated Densilog Caliper Neutron	C DLC N
Compensated Densilog Caliper G/R Neutron Compensated Densilog Caliper Minilog	C DLC C DL GRN C M
Conductivity Derived Porosity	CDP
Corgun	CG
Densilog Caliper Gamma Ray Log	DLC GR
Depth Determination	DD
Directional Survey	DS
Dual Induction Focused Log	DIFL
Dual Induction Focused Log Gamma Ray	DIFL GR
Electrolog	EL
Focused Diplog	F DIP
Formation Tester	FT
4 Arm High Resolution Diplog	RH DIP
Frac Log	FRAC L
Frac Log-Gamma Ray	FRAC-GR
Gamma Ray Cased Hole	GR CH
Gamma Ray/Dual Caliper	GR/D CALIPER
Gamma Ray-Open Hole	G/R OH
Gamma Ray Neutron-Cased Hole	GRN CH
Gamma Ray Neutron-Open Hole	GR/N OH
Geophone	GEO
Induction Electrolog	IEL
Induction Electrolog Gamma Ray	IEL-GR
Induction Electrolog Neutron	IEL-N
Induction Electrolog Gamma Ray Neutron	IEL-GRN
Induction Log	IL
Induction Log-Gamma Ray	IL-GR
Induction Log-Neutron	IL-N
Laterolog	LL
Laterolog-Gamma Ray	LL-GR
Laterolog-Neutron	LL-N
Laterolog-Gamma Ray-Neutron	LL-GRN
Microlaterolog-Caliper	MLLC
Minilog Caliper	ML-C
Minilog Caliper Gamma Ray	ML-C-GR
Movable Oil Plot	MOP
Neutron Cased Hole	N CH

Neutron Cement Log	N CL
Neutron Open Hole	N OH
Neutron Lifetime	NLL
Neutron Lifetime Gamma Ray	NLL GR
Neutron Lifetime Neutron	NLL N
Neutron Lifetime G/R-Neutron	NLL GRN
Neutron Lifetime CBL	NLL CBL
Neutron Lifetime G/R-CBL	NLL GR CBL
Neutron Lifetime CBL-Neutron	NL CB N
Neutron Lifetime CBL-G/R-Neutron	NLL CBL GR
Nuclear Flolog	NFL
Nuclear Flolog-Gamma Ray	NFL GR
Nuclear Flolog-Gamma Ray Neutron	NFL GRN
Nuclear Flolog-Neutron	NFL N
Nuclear Cement Log	NCL
Perforating Control	PFC
PFC Gamma Ray	PFC GR
PFC Neutron	PFC N
Photon	PL
Proximity Minilog	PROX-MLC
Sidewall Neutron	SWN
Sidewall Neutron-Gamma Ray	SWN GR
Temperature Differential	DTL
Temperature-Gamma Ray-Neutron	T GRN
Temperature Log	TL
Temperature Log-Gamma Ray	T GR
Temperature-Neutron	T N
Total Time Integrator	TTI
Tracer Log	TL
Tracer Log-Neutron	TLN
Tracer Material	TM
Tracer Placement with Dump Bailer	TV DB
Tricore	TCS

GO INTERNATIONAL

Caliper	CALP
Cement Bond Log	CBL
Differential Temperature	DIF-T
Gamma Ray	GR
Gamma Ray Neutron	GR-N
Neutron	N
Temperature Log	T

SCHLUMBERGER WELL SERVICES

Amplitude Logging	A-BHC
Bore Hole Compensated	BHC
BHC Sonic Logging	BHC
BHC Sonic-Gamma Ray Logging	BHC-GR
BHC Sonic-Variable Density	BHC-VD
Bridge Plug Service	BP
Borehole Televiewer	TVT
Caliper Logging	CAL
Casing Cutter Service	SCE-CC
Cement Bond Logging	CBL
Cement Bond-Gamma Ray Logging	CBL-GR
Cement Bond-Gamma Ray Neutron	CBL-GRN
Cement Bond-Neutron	CBL-N
Cement Bond-Variable Density Logging	CBL-VD
Cement Dump Bailer Service	DB
Computer Processed Interpretation	MCT
Continuous Directional Survey	CDR
Continuous Flowmeter	CFM, PFM
Customer Instrument Service	ICS
Data Transmission	TRD
Density Log	DENL
Depth Determinations	DD
Diamond Core Slicer	SS
Dipmeter	DIPM
Dipmeter-Digital	HDT-D
Directional Survey	DS
Dual Induction-Laterologging	DIL
Electric Logging	ES
Formation Density Logging	FDC
Formation Density-Gamma Ray Logging	FDC-GR
Formation Testing	FT
Gamma Ray Logging	GR
Gamma Ray-Neutron Logging	GRN
Gamma Ray-Sonic Logging	GRS
Gradiomanometer	GM
High-Resolution Thermometer	HRT
Induction-Electron Logging	I-ES
Induction-Gamma Ray Logging	I-GR
Junk Catcher	JB
Log Overlays	OL
Magnetic Taping	TPG

Microlog	ML
Neutron Logging	NL
Orienting Perforating Service	OPR
Perforating-Ceramic DPC	SCE
Perforating-Depth Control	PDC
Perforating-Expendable Shaped Charge	SCE
Perforating-Hyper Jet	SCH
Perforating-Hyper Scallop	SPH
Pressure Control	PC
Production Combination Tool Logging	PCT
Production Packer Service	PPS
Proximity-Microlog	ML
Radioactive Tracer Logging	RTP
Rwa Logging	FAL
Salt Dome Profiling	ES-ULS
Schlumberger	Schl.
Seismic Reference Service	SRS
Sidewall Coring	CST
SNP Neutron Logging	SNP
SNP Neutron-Gamma Ray Logging	SNP-GR
Synergetic Log Systems	MCT
Temperature Logging	T
Temperature-Gamma Ray Logging	T-GR
Thermal Decay Logging	TDT
Thru-Tubing Caliper	C-C
Tubing, Cutter Service	SEC-CC
Variable Density Logging	BHC-VD
Variable Density-Gamma Ray Logging	VD-GR

WELEX

Analog Computer Service	An Cpt. Ser
Caliper	Cal
Compensated Acoustic Velocity Log	Com AVL
Compensated Acoustic Velocity Log-Gamma Ray	Com AVL-G
Compensated Acoustic Velocity Log-Neutron	Com AVL-N
Compensated Density	Com Den
Compensated Density Gamma Ray	Com Den-GR
Computer Analyzed Log	CAL
Contact Caliper	Cont

Continuous Drift	Con Dr.
Density	Den
Density Gamma Ray	Den-G
Depth Determination	DeDet
Digital Tape Recording	Dgt Tp Rec
Dip Log Digital Recording	Dgt Dip Rec
Drift	Dr
Drill Pipe Electric Log	DPL
Electric Log	EL
Electro-Magnetic Corrosion Detector	Cor Det
Fluid Travel Log	FTrL
Formation Tester	FT
FoRxo Caliper	FoRxo
Frac-Finder Micro-Seismogram	FF-MSG
Frac-Finder Micro-Seismogram Gamma	FF-MSG-G
Frac-Finder Micro-Seismogram Neutron	FF-MSG-N
Gamma Guard	G-Grd
Gamma Ray	GR
Gamma Ray Depth Control	GRDC
Guard	Grd
High Temperature Equipment	HTEq
Induction Electric Gamma Ray	IEL-G
Induction Electric Neutron	IEL-N
Induction Electric	IEL
Induction Gamma Ray	Ind-G
Induction Gamma	Ind G
Micro-Seismogram Log, Cased Hole	MSG-CBL
Micro-Seismogram Gamma Collar Log, Cased	MSG-CBL-G
Micro-Seismogram Neutron Collar Log, Cased	MSG-CBL-N
Neutron Log	NL
Neutron Depth Control	NDC
Precision Temperature	Pr Temp
Radiation Guard	R/A Grd
Radioactive Tracer	R/A Tra
Resistivity Dip	Dip
Sidewall Coring	SWC
Sidewall Neutron	SWN
Sidewall Neutron-Gamma Ray	SWN-G
Simultaneous Gamma Ray-Neutron	GRN
Special Instrument Service	Sp Inst Ser
True Vertical Depth	TVD

MISCELLANEOUS

Acoustic Amplitude	**AAL**
Acoustic Cement Bond G/R Neutron	**CBL GRN**
Acoustic Cement Bond Neutron	**CBL N**
Acoustic Cement Bond	**CBL**
Acoustic Fracture Identification	**AFI**
Acoustic or Acoustilog	**SL**
Acoustic Parameter	**ACP**
Acoustic Parameter-Depth	**AC PAR D**
Acoustic Parameter-Logging	**AC PAR L**
Acoustic Parameter-16mm Scope	**AC PAR 16**
Acoustic Scope Picture	**ASL**
Acoustic Signature	**AC SIGN**
Acoustic Velocity	**AVL**
Acoustilog	**ALC**
Acoustilog Caliper-Gamma Ray	**ALC-GR**
Acoustilog Caliper-Gamma Ray-Neutron	**ALC-GRN**
Acoustilog Caliper-Neutron	**ALC-N**
Amplitude	**AMP**
Amplitude Sonic	**ASL**
Atlantic Chlorinlog	**A CHL**
Audio Logging	**AUD**
BHC Acoustilog Caliper	**BHC ALC**
BHC Acoustilog Caliper (Thru Casing)	**BHC AL TC**
BHC Acoustilog Caliper G/R Neutron	**BHC ALC GRN**
BHC Acoustilog Caliper G/R Neutron (Thru Casing)	**BHC AL GRN TC**
BHC Acoustilog Caliper Gamma Ray	**BHC ALC GR**
BHC Acoustilog Caliper Neutron	**BHC ALC N**
Borehole Audio Tracer Survey	**BATS**
Borehole Compensated	**BHC**
Borehole Compensated Sonic	**BHCS**
Borehole Geometry Log	**BGT**

Borehole Televiewer	**BTL**
Bulk Density	**BLKD**
Caliper	**CALP, CL**
Caliper Analysis	**CALA**
Caliper Curve	**CALC**
Carbon-Oxygen	**C O**
Casing Collar	**CCL**
Casing Inspection/Electro-Magnetic-Detector	**CI**
Casing Potential Profile	**CPP**
Cement Bond	**CBND**
Cement Evaluation	**CEL**
Cement Top Location	**CTL**
Cemotop	**CTL**
Channel Survey	**CHNL**
Channelmaster	**CML**
Channelmaster-Neutron	**CML N**
Chloride	**CL**
Chloride Detection	**CLDL**
Chlorinlog-Gamma Ray	**CHL GR**
Collar/Collar Correlation	**PDS**
Compensated Acoustic Velocity	**CAVL**
Compensated Densilog Caliper	**CDCL**
Compensated Densilog Caliper-G/R Neutron	**CDLC GRN**
Compensated Densilog Caliper-Gamma Ray	**CDLC GR**
Compensated Densilog Caliper-Minilog	**CDLC M**
Compensated Densilog Caliper-Neutron	**CDLC N**
Compensated Density Caliper	**CDC**
Compensated Density Log	**CDL**
Compensated Formation Density	**CFD**
Compensated Formation Density Caliper	**CFDC**
Compensated Gamma	**CG**
Compensated Neutron Density	**CNDL**
Compensated Neutron Log	**CNL**
Compensated Neutron Log Porosity	**CNLP**

Borehole Televiewer	**BTL**
Bulk Density	**BLKD**
Caliper	**CALP, CL**
Caliper Analysis	**CALA**
Caliper Curve	**CALC**
Carbon-Oxygen	**C O**
Casing Collar	**CCL**
Casing Inspection/Electro- Magnetic-Detector	**CI**
Casing Potential Profile	**CPP**
Cement Bond	**CBND**
Cement Evaluation	**CEL**
Cement Top Location	**CTL**
Cemotop	**CTL**
Channel Survey	**CHNL**
Channelmaster	**CML**
Channelmaster-Neutron	**CML N**
Chloride	**CL**
Chloride Detection	**CLDL**
Chlorinlog-Gamma Ray	**CHL GR**
Collar/Collar Correlation	**PDS**
Compensated Acoustic Velocity	**CAVL**
Compensated Densilog Caliper	**CDCL**
Compensated Densilog Caliper- G/R Neutron	**CDLC GRN**
Compensated Densilog Caliper- Gamma Ray	**CDCL GR**
Compensated Densilog Caliper- Minilog	**CDLC M**
Compensated Densilog Caliper- Neutron	**CDLC N**
Compensated Density Caliper	**CDC**
Compensated Density Log	**CDL**
Compensated Formation Density	**CFD**
Compensated Formation Density Caliper	**CFDC**
Compensated Gamma	**CG**
Compensated Neutron Density	**CNDL**
Compensated Neutron Log	**CNL**
Compensated Neutron Log Porosity	**CNLP**

Dual Induction Gamma Log	**DIGL**
Dual Induction Lateral/Dual Induction Focus	**DILL**
Dual Induction Log	**DIL**
Dual Induction SFL	**DIL**
Dual Induction Spherically Focused	**DISF**
Dual Laterolog/Microspherically Focused	**DLL/MSFL**
Dual Laterolog	**DLL**
Dual Porosity Compensated Neutron	**DNL**
Dual Resistivity Induction Log	**DRI**
Dual Sand	**DUSD**
Dual Spacing Log	**DSL**
Electrical	**ES**
Electrolog	**EL**
Electromagnetic Propagation	**EPT**
Epithermal Neutron	**ETN**
Experimental	**E, XPTL**
Flowmeter	**FLO**
Fluid Travel	**FLO**
Fluid Travel Log/Fluid Entry Survey	**FLTR**
Focus	**FOCL**
Focused Diplog	**F DIP**
Formation Analysis	**FAL**
Formation Density	**FD**
Formation Density Caliper	**FDC**
Formation Factor	**FMF**
Formation Tester	**FT**
Forxo	**MLL**
4 Arm High Resolution Diplog	**R H DIP**
Frac Log	**FRAC L**
Frac Log-Gamma Ray Log	**FRAC-GR**
Fracture Finder	**ASL**
Fracture Finder/Failure ID	**FF**
Fracture Identification Log	**FIL**
Full Bore Flowmeter	**FB FM**
Gamma Compensated Density	**GCD**

Gamma Gamma	**GG**
Gamma Gamma Density	**GGD**
Gamma Gamma Log	**GGL**
Gamma Guard	**GCRD**
Gamma Guard EL	**GGRD**
Gamma Ray	**GR**
Gamma Ray - Multi-Spaced Neutron Log	**CR-MSN**
Gamma Ray - Neutron	**GRNL**
Gamma Ray - Neutron Log	**GR-N**
Gamma Ray - Sonic	**GRSL**
Gamma Ray - Tracer Survey	**GRTS**
Gamma Ray Cased Hole	**GR CH**
Gamma Ray Depth Control	**GRDC**
Gamma Ray Neutron	**GRN**
Gamma Ray Neutron-Cased Hole	**GRN CH**
Gamma Ray Neutron-Open Hole	**GR/N OH**
Gamma Ray-Open Hole	**G/R OH**
Gamma Ray/Dual Caliper	**GR/D CALIPER**
Gamma Spectrometry	**GST**
Gas Detection	**GASD**
Geophone	**GEO**
Gradiomanometer	**GRMR**
Gravity	**GRAV**
Guard	**GRDL**
Gyro Survey	**GYRO**
High Resolution Dipmeter	**HRD**
Hydrocarbon or Gas Detection	**GT**
Hydrocarbon or Gas Detection	**HCDS**
Inclination	**INCL**
Induction	**IL**
Induction-Electric Log	**I-EL**
Induction-Lateral	**ILL**
Induction-Laterolog	**I-LL**
Induction Electrolog	**IEL**
Induction Electrolog-Gamma Ray	**IEL-GR**
Induction Electrolog-Gamma Ray Neutron	**IEL-GRN**
Induction Electrolog-Neutron	**IEL-N**

Induction Spherically Focused	**ISF**
Induction-Electric or Induction-Electro	**IES**
Isotron	**I, ISOL**
Laser Log	**LASR**
Lateral	**LATL**
Laterolog-Gamma Ray-Neutron Log	**LL GR-N**
Lifetime Log	**LL**
Limestone Log	**LSL**
Limestone Device	**LI**
Liquid Isotope Injector	**LII**
Litho-Density	**LDT**
Lithology	**LITH**
Logger's Total Depth	**LTD**
Long-Spaced Sonic	**LSS**
Lost Circulation	**LS**
Manometer	**MAN**
Microlog	**ML, MICL**
Microlateral	**MLAT**
Microseismogram	**MSMG**
Microsonic Gamma Ray	**MSG**
Microspherically Focused Log	**MSFL**
Microsurvey	**MS, MICS**
Mini	**ML, MINL**
Minifocus	**MLL, MINF**
Mobile Picture	**MP**
Mono Electric	**MONO**
Mud Log/Focus Log	**MUD**
Multi-Shot Survey	**MSS**
Natural Gamma Ray Spectroscopy	**NGS**
Neutron	**N, NEUT**
Neutron Collar Log	**NCL**
Neutron Formation Density	**NFD**
Neutron Lifetime	**NLL**
Nuclear	**NUCR**
Nuclear Flow	**FLO**
Nuclear Magnetism	**NML, NMAGL**

Perforated Log	**PERF**
Perforating Depth Control	**PDC**
Perforating Formation Collar	**PFC**
Permalog	**PL**
Permeability Spinner Survey	**PSS, PRMS**
Photo	**PHOT**
Photoclinometer	**PHCL**
Pipe Analysis Log	**PAL**
Pipe Recovery Log	**PRL**
Porosity	**POR**
Proximity	**PROX**
Proximity Log	**PROXL**
Proximity-Microlog	**PML**
Radioactive Tracer	**RAT, RTRS**
Refracture	**"REFR**
Repeat Formation Tester	**RFT**
Resistivity	**RES**
Resistivity Water Apparent	**RWA**
Salinity	**CL**
Saraband	**"SBND**
Scattered Gamma Ray	**CTL**
Scope Picture Analysis	**SPA**
Section Gauge	**CAL**
Seismic Reference Survey/ Neutron	**SRSN**
Seismic Velocity Survey	**SVS**
Shear Amplitude	**SA**
Sidewall Frac Log	**SFL**
Sidewall Neutron	**SN**
Sidewall Neutron Porosity	**SNP**
Sidewall Sampler	**CST**
Sonic Caliper Log	**SCL**
Sonic Log	**SL, SONL**
Sonic Seismogram	**SSMG**
Spectral	**SPCT**
Spherical	**SPH**
Spontaneous Potential	**SP**
Strata	**STRTS-tructural Exploration**

Structural Exploration	**SE**
Synergetic	**SYGT**
Televiewer	**TV**
Temperature	**HRT**
Temperature Difference Log	**TDL**
Temperature Survey Log	**TMPL**
Thermal Decay Time	**TDT**
Three Dimensional	**3D**
Time Log	**TIME**
Tracer Survey	**TRCR**
Ultralong Spacing Electric Log	**ULSEL**
Uranium	**URAN**
Variable Density	**VD**
Velocity	**VEL**
Velocity Seismic Profile	**VSP**
Velocity Survey	**VRS**
Velocity Survey Profile	**VSP**
Viscosity	**VISC**
Water Location Survey	**WLS**
Wave Form Digitizing	**BHC-WFD**
Wave Form Logging	**BHC-WFL**
Well Seismic	**WST**
X-Y Caliper	**XYCL**

PIPE COATING TERMINOLOGY
AND DEFINITIONS

anode	Corrosion prevention device
C.P.	Cathodic protection or Corrision Protection
dope	Pipe coating
dresser	Mechanical coupling used to join joints or lengths or pipe rather than threading or welding
FB	Flat-bottom mill or shoes
granny rag	Type of coating or method of coating a pipeline in the field rather than factory applied coating
holiday	Hole in the protective coating of a steel pipeline in the field
hot spot	Corrosive area located along the length of a pipeline; usually a wet bog, marsh, or bentonite area
I.P.	Intermediate pressure pipeline
IWRC	Independent wire rope center
jeep	Same as *holiday*
jeeper	Electronic device or instrument used to detect holes (holidays) in the steel pipeline protective coating
overbend	High spot in a pipeline usually installed by field-bending a pipeline joint
P/C or P/W	"Painted and coated" or "painted and wrapped" pipelines—steel pipe with protective external coating of one of several different types

pig	Pipeline cleaning and measuring tool
pig catcher	Used to remove pipeline pig
pig launcher	Used to insert pipeline pig
stub	Length of small-diameter distribution pipeline from the main line to the customer's property or meter location
thin film	Type of epoxy coating for pipeline coating rather than threading or welding
tube turn	Prefabricated piece of pipeline (allows change of direction of a pipeline without field bending the pipeline)
sag	Low spot in a pipeline usually installed by field-bending a pipeline joint
WB	Wavy bottom mill or shoes
wrap	Same as dope; protective coating on a steel pipeline
XTC	Extra-coat protective pipeline coating made of polyethylene or polypropylene material

COMPANIES AND ASSOCIATIONS
U.S. and Canada

AAODC	See IADC
AAPG	American Assocation of Petroleum Geologists
AAPL	American Association of Petroleum Landmen
AAR	Associaton of American Railroads
ABSORB	Alaska Beaufort Sear Oil Spill Response Body
ACMP	Alaska Coastal Management Program
ACS	American Chemical Society
ADDA	Association of Desk and Derrick Clubs of North America
AEC	Atomic Energy Commission
AECRB	Alberta Energy Conservation Resources Board
AGA	American Gas Association
AGI	American Geological Institute
AGTL	Alberta Gas Trunkline Co., Ltd.
AGU	American Geophysical Union
AIChE	American Institute of Chemical Engineers
AIME	American Institute of Mining, Metallurgical and Petroleum Engineers
AISI	American Iron and Steel Institute
ALCOA	Aluminum Company of American
ANSI	American National Standards Institute
AOAC	Association of Official Agricultural Chemists

AOCS	American Oil Chemists Society
AOGA	Alaskan Oil & Gas Association
AOPL	Association of Oil Pipe Lines
AOSC	Association of Oilwell Servicing Contractors
AP&VMA	American Paint & Varnish Manufacturers Assn.
APHA	American Public Health Association
API	American Petroleum Institute
APRA	American Petroleum Refiners Association
APW	Association of Petroleum Writers
ARCO	Atlantic Richfield Co.
ARKLA	Arkansas Louisiana Gas Co.
ASA	American Standards Assoc.
ASCE	American Society of Civil Engineers
ASHRAE	American Society of Heating, Refrigerating, and Air-Conditioning Engineers, Inc.
ASLE	American Society of Lubricating Engineers
ASME	American Society of Mechanical Engineers
ASPG	American Society of Professional Geologists
ASSE	American Society of Safety Engineers
ASTM	American Society for Testing Materials
AWS	American Welding Society
AWWA	American Water Works Association
BLM	Bureau of Land Management

BLS	Bureau of Labor Statistics
BP	British Petroleum
BuMines	Bureau of Mines, U.S. Department of the Interior
CAGC	A combine: Continental Oil Co., Atlantic Richfield Co., Getty Oil Co., and Cities Service Oil Co.
CAODS	Canadian Association of Oilwell Drilling Contractors
CCCOP	Conservation Committee of California Oil Producers
CDS	Canadian Development Corp.
CFR	Coordinating Fuel Research Committee
CFRC	Coordinating Fuel Research Committee
CGA	Clean Gulf Associates
CGTC	Columbia Gas Transmission Co.
CL&F	Continental Land and Fur Co.
COE	Corps of Engineers
CONOCO	Continental Oil Co.
CORCO	Commonwealth Oil Refining Co., Inc.
CORS	Canadian Operational Research Society
CPA	Canadian Petroleum Association
CRC	Coordinating Research Council, Inc.
DOE	Department of Energy
DOT	Department of Transportation
DNR	Department of Natural Resources

DRAPR	Delaware River Area Petroleum Refineries
Drssr., DA	Dresser Atlas
EMR	Department of Energy, Mines, and Resources (Canada)
EPA	Environmental Protection Agency
ERCB	Energy Resource Conservation Board (Alberta, Canada)
FAA	Federal Aviation Agency
FCC	Federal Communications Commission
FPC	Federal Power Commission
FSIWA	Federation of Sewage and Industrial Wastes Assn.
FTC	Federal Trade Commission
GA	Canadian Gas Association
GAMA	Gas Appliance Manufacturers Association
GE	General Electric Company
GM	General Motors
GNEC	General Nuclear Engineering Co.
GR&DC	Gulf Research and Development Company
IADC	International Association of Drilling Contractors (formerly AAODC)
IAE	Institute of Automotive Engineers
ICC	Interstate Commerce Commission
IEEE	Institute of Electrical and Electronics Engineers
IGT	Institute of Gas Technology
INGAA	Independent Natural Gas Association of America

IOCA	Independent Oil Compounders Association
IOCC	Interstate Oil Compact Commission
IOSA	International Oil Scouts Association
IP	Institute of Petroleum
IPAA	Independent Petroleum Association of America
IPAC	Independent Petroleum Association of Canada
IPE	International Petroleum Exposition
IPP/L	Interprovincial Pipe Line Co.
IRAA	Independent Refiners Association of America
ISA	Instrument Society of America
JCUMWE	Joint Committee on Uniformity of Methods of Water Examinaton
KERMAC	Kerr-McGee Corp.
KIOGA	Kansas Independent Oil and Gas Association
LL&E	Louisiana Land & Exploration Co.
MIOP	Mandatory Oil Import Program
MMS	Minerals Management Service
NACE	National Association of Corrosion
NACOPS	National Advisory Committee on Petroleum Statistics (Canada)
NAS	National Academy of Science
NASA	National Aeronautical and Space Administration

NEB	National Energy Board (Canada)
NEMA	National Electrical Manufacturers Association
NEPA	National Environmental Policy Act of 1969
NGPA	Natural Gas Processor Association
NGPSA	Natural Gas Processors Suppliers Association
NLGI	National Lubricating Grease Institute
NLPGA	National Liquefied Petroleum Gas Association
NLRB	National Labor Relations Board
NOFI	National Oil Fuel Institute
NOIA	National Ocean Industries Association
NOJC	National Oil Jobbers Council
NOMADS	National Oil-Equipment Manufacturers and Delegates Society
NPC	National Petroleum Council
NPDES	National Pollution Discharge Elimination System
NPR	Naval Petroleum Reserve
NPRA	Naval Petroleum Reserve, Alaska
NPRA	National Petroleum Refiners Association
NRC	Nuclear Regulatory Commission
NSF	National Science Foundation
OCR	Office of Coal Research
OEP	Office of Emergency Preparedness

OIA	Oil Import Administration
OIAB	Oil Import Appeals Board
OIC	Oil Information Committee
OIPA	Oklahoma Independent Petroleum Association
OOC	Offshore Operators Committee
OPC	Oil Policy Committee
ORIAC	Oil Refining Industry Action Committee
ORSANCO	Ohio River Valley Water Sanitation Commission
OXY	Occidental Petroleum Corp.
PAD	Petroleum Administration for Defense
PESA	Petroleum Equipment Suppliers Association
PETCO	Petroleum Corporation of Texas
PGCOA	Pennsylvania Grade Crude Oil Association
PIEA	Petroleum Industry Electrical Association
Plato	Pennzoil Louisiana and Texas Offshore
PLCA	Pipe Line Contractors Association
POGO	Pennzoil Offshore Gas Operators
PPI	Plastic Pipe Institute
PPROA	Panhandle Producers and Royalty Owners Association
RMOGA	Rocky Mountain Oil and Gas Association
R-PAT	Regional Petroleum Association
RPI	Research Planning Institute
RTL	Refinery Technology Laboratory

SACROC	Scurry Area Canyon Reef Operators Committee
SAE	Society of Automotive Engineers
Schl., Sj	Schlumberger
SEG	Society of Exloration Geophysicists
SEPM	Society of Economic Paleontologists and Mineralogists
SGA	Southern Gas Associaton
SLAM	A combine: Signal Oil and Gas Co., Louisiana Land & Exploration Co., Amerada Hess Corp., and Marathon Oil Co.
SOCAL	Standard Oil Company of California
SOHIO	Standard Oil Co. of Ohio
SPE	Society of Petroleum Engineers of AIME
SPEE	Society of Petroleum Evaluation Engineers
SPWLA	Society of Professional Well Log Analysts
STATCAN	Statistics Canada ex Dominion, Bureau of Statistics (DBS)
TAPS	Trans-Alaska Pipeline Systems
TCP	Trans-Canada Pipe Lines Ltd.
TETCO	Texas Eastern Transmission Corp.
TGT	Tennessee Gas Transmission Co.
THUMS	A combine: Texaco, Inc., Humble Oil & Refining Co., Union Oil Co. of California, Mobil Oil Corp., and Shell Oil Co.

TIPRO	Texas Independent Producers and Royalty Owners Association
TRANSCO	Transcontinental Gas Pipe Line Corp.
UOCO	Union Oil Company
UOP	Universal Oil Products Company
USGS	United States Geological Survey
USP	United States Pharmocopoeia
WeCTOGA	West Central Texas Oil and Gas Association
WOGA	Western Oil & Gas Association
WPC	World Petroleum Congress
Wx	Welex

COMPANIES AND ASSOCIATIONS
Outside the U.S. and Canada

AAOC	American Asiatic Oil Corp. (Philippines)
ABCD	Asfalti Bitumi Cementi Derivati, S.A. (Italy)
ACNA	Aziende Colori Nazionali Affini (Italy)
A.C.P.H.A.	Association Cooperative pour la Recherche et l'Exploration des Hydrocarbures en Algerie (Algeria)
ADCO-HH	African Drilling Co.-H. Hamouda (Libya)
AGIP S.p.A.	Azienda Generale Italiana Petroli S.p.A. (Italy)
A.H.I. BAU	Allegemeine Hoch-und Ingenieurbau AG (Germany)
AIOC	American International Oil Co. (U.S.A.)
AITASA	Aguas Industriales de Tarragona, S.A. (Spain)
AK CHEMI	GmbH & Co. KG–subsidiary of Associated Octel, Ltd., London, Eng (Germany)
AKU	Algemene Kunstzijde Unie, N.V. (Netherlands)
ATAS	Anadolu Tastiyehanesi A.S. (Turkey)
AUXERAP	Societe Auxiliaire de la Regie Autonome des Petroles (France)
AZOLACQ	Societe Chimique d'Engrais et de Produits de Synthese (France)
BAPCO	Bahrain Petroleum Co. Ltd. (Bahrain)

BASF	Badische Anilin & Soda-Fabrik AB (Germany)
BASUCOL	Barranquilla Supply & Co. (Colombia)
B.I.P.M.	Bataafse Internationale Petroleum Mij. N.V. (Netherlands)
BOGOC	Bolivian Gulf Oil Co. (Bolivia)
BORCO	Bahamas Oil Refining Co. (Bahamas)
BP	British Petroleum Co., Ltd. (England)
BRGG	Bureau de Recherches Geologiques et Geophysique (France)
BRGM	Bureau de Recherches Geologique et Minieres (France)
BRIGITTA	Gewerkschaft Brigitta (Germany)
BRP	Bureau de Recherche de Petrole (France)
BRPM	Bureau de Recherches et de Participations Mineres (Morocco)
BSI	British Standards Institute
CALSPAIN	California Oil Co. of Spain (Spain)
CALTEX	Various affiliates of Texaco Inc. and Std. of Calif.
CALVO SOTELO	Empresa Nacional Calvo Sotelo (Spain)
CAMEL	Campagnie Algerienne du Methane Liquide (France, Algeria)
CAMPSA	Compania Arrendataria del Monopolio de Petroleos, S.A. (Spain)

CAPAG	Enterprise Moderne de Canalisations Petrolieres, Aquiferes et Gazieres (France)
CARBESA	Carbon Black Espanola, S.A. (Conoco affiliate)
CAREP	Compagnie Algerienne de Recherche et d'Exploitation Petrolieres (Algiers)
CCC	Compania Carbonos Coloidais (Brazil)
C.E.C.A.	Carbonisation et Charbons Actifs S.A. (France)
CEICO	Central Espanol Ingenieria y Control S.A. (Spain)
CEL	Central European Pipeline (Germany)
CEOA	Centre Europe de'Exploitation de I'OTAN (France)
CEP	Compagnie D'Exploration Petroliere (France)
CEPSA	Compania Espanola de Petroleos, S.A. (Spain)
CETRA	Compagnie Europeanne de Canalisations et de Travaux (France)
CFEM	Compagnie Francaise d'Enterprises Metalliques (France)
CFM	Compagnie Francaise du Methane (France)
CFMK	Compagnie Ferguson Morrison-Knudison (France)
CFP	Compagnie Francaise du Petroles (France)
CFPA	Compagnie Francaise des Petroles (Algeria) (France)

CFPS	Compagnie Francaise de Prospection Sismique (France)
CFR	Compagnie Francaise de Raffinage (France)
CGG	Compagnie Generale de Geophysique (France, Australia, Singapore)
CIAGO	N.V. Chemische Industrie aku-Goodrich (Netherlands)
CIEPSA	Compania de Investigacion y Explotaciones Petroliferas, S.A. (Spain)
CIM	Compagnie Industrielle Maritime (France)
CIMI	Compania Italiana Montaggi Industriali S.p.a. (Italy)
CINSA	Compania Insular del Nitrogena, S.A. (Spain)
CIPAO	Compagnie Industrielle des Petroles de I'A.O. (France)
CIPSA	Compania Iberica de Prospecciones, S.A. (Spain)
CIRES	Compania Industrial de Resinas Sinteticas (Portugal)
CLASA	Carburanti Librificanti Affini S.p.A. (Italy)
CMF	Construzioni Metalliche Finsider S.p.A. (Italy)
COCHIME	Compagnie Chimique de la Meterranee (France)
CODI	Colombianos Distribuidores de Combustibles S.A. (Colombia)
COFIREP	Compagnie Financiere de Recherches Petrolieres (France)

COFOR	Compagnie Generale de Forages (France)
COLCITO	Colombia-Cities Service Petroleum Corp. (Colombia)
COLPET	Colombian Petroleum Co. (Colombia)
COMEX	Compagnie Maritime d'Expertises (France)
CONSPAIN	Continental Oil Co. of Spain (Spain), Conoco Espanola S.A. (Spain)
COPAREX	Compagnie de Participations, de Recherches et d'Exploitations Petrolieres (France)
COPE	Compagnie Orientale des Petroles d'Egypte (Egypt)
COPEBRAS	Compania Petroquimica Brasileira (Brazil)
COPEFA	Compagnie des Petroles France-Afrique (France)
COPETAO	Compagnie des Petroles Total (Afrique Quest) (France)
COPETMA	Compagnie les Petroles Total (Madagascar)
COPISA	Compania Petrolifera Iberica, Sociedad Anonima (Spain)
COPSEP	Compagnie des Petroles du Sud est Parisien (France)
C.O.R.I.	Compania Richerche Idrocarburi S.p.A. (Italy)
COS	Coordinated Oil Services (France)
CPA	Compagnie des Petroles d'Algeria (Algeria)

CPC	Chinese Petroleum Corporation, Taiwan, China
CPTL	Compagnie des Petroles Total (Libye) (France)
CRAN	Compagnie de Raffinage en Afrique du Nord (Algeria)
CREPS	Compagnie de Recherches et d'Exploitation de Petrole au Sahara (Algeria)
CRR	Compagnie Rhenane de Raffinage (France)
CSRPG	Chambre Syndicale de la Recherche et de la Production du Petrole et du Gaz Naturel (France)
CTIP	Compania Tecnica Industrie Petroli S.p.a. (Italy)
CVP	Corporacion Venezolano del Petroleo (Venezuela)
DCEA	Direction Centrale des Essences des Armees (France)
DEA	Deutsche Erdol-Aktiengesellschaft (Germany)
DEMINEX	Deutsch Erdolversorgungs-gesellschaft mbH (Germany) (Trinidad)
DIAMEX	Diamond Chemicals de Mexico, S.A. de C.V. (Mexico)
DICA	Direction des Carburants (France)
DICA	Distilleria Italiana Carburanti Affini (Italy)
DITTA	Macchia Averardo (Italy)
DUPETCO	Dubai Petroleum Company (Trucial States)

E.A.O.R.	East African Oil Refineries, Ltd. (Kenya)
ECF	Essences et Carburants de France (France)
ECOPETROL	Empresa Colombiana de Petroleos (Colombia)
EGTA	Enterprises et Grands Travaux de l'Atlantique (France)
ELF-ERAP	Enterprise de Recherches et d'Activites Petrolieres (France)
ELF-U.I.P.	Elf Union Industrielle des Petroles (France)
SLF—SPAFE	Elf des Petroles D'Afrique Equatoriale (France)
ELGI	M. All. Eotvos Lorand Geofizikai Intezet (Hungary)
ENAP	Empresa Nacional del Petroleo (Chile)
ENCAL	Engenheiros Consultores Associados S.A. (Brazil)
ENCASO	Empresa Nacional Calvo Sotelo de Combustibles Liquidos y Lubricantes, S.A. (Spain)
ENGEBRAS	Engenharia Especializada Brasileira, S.A. (Brazil) (Venezuela)
ENI	Ente Nazionale Idrocarburi (Italy)
ENPASA	Empresa Nacional de Petroleos de Aragon, S.A. (Spain)
ENPENSA	Empresa Nacional de Petroleos de Navarra, S.A. (Spain)
ERAP	Enterprise de Recherches et d'Activites Petrolieres (France)

ESSAF	Esso Standard Societe Anonyme Francaise (France)
ESSOPETROL	Esso Petroleos Espanoles, S.A. (Spain)
ESSO REP	Societe Esso de Recherches et Exploitation Petrolieres (France)
E.T.P.M.	Societe Entrepose G.T.M. pour les Travaux Petroliers Maritimes (France)
EURAFREP	Societe de Recherches et d'Exploitation de Petrole (France)
FERTIBERIA	Fertilizantes de Iberia, S.A. (Spain)
FFC	Federation Francaise des Carburants (France)
FINAREP	Societe Financiere des Petroles (France)
FOREX	Societe Forex Forages et Exploitation Petrolieres (United Kingdom)
FRANCAREP	Compagnie Franco-Africaine de Recherches Petrolieres (France)
FRAP	Societe de Construction de Feeders, Raffineries, Adductions d'Eau et Pipe-Lines (France)
FRISIA	Erdolwerke Frisia A.G. (Germany)
GARRONE	Garrone (Dott. Edoardo) Raffineria Petroli S.a.S. (Italy)
GBAG	Gelsenberg Benzin (Germany)
GESCO	General Engineering Services (Colombia)

GHAIP	Ghanian Italian Petroleum Co., Ltd. (Ghana)
GO INTL	GO International, Inc.
GPC	The General Petroleum Co. (Egypt)
G.T.M.	Les Grands Travaux de Marseille (France)
HELIECUADOR	Helicopteros Nacionales S.A. (Ecuador)
HDC	Hoecsht Dyes & Chemicals Ltd. (India)
HIDECA	Hidrocarburos y Derivados C.A. (Brazil, Uraguay & Venezuela)
HIP	Hemijska Industrija Pancevo (Yugoslavia)
H.I.S.A.	Herramientas Interamericanas, S.A. de C.V. (Mexico)
HISPANOIL	Hispanica de Petroleos, S.A., (Spain)
HOC	Hindustan Organic Chemicals Ltd. (India)
HYLSA	Hojalata y Lamina, S.A. (Mexico)
IAP	Institut Algerien du Petrole (Algeria)
ICI	Imperial Chemical Industries Ltd. (England)
ICIANZ	Imperial Chemical Industries of Australia & New Zealand Ltd. (Australia, New Zealand)
ICIP	Industrie Chimiche Italiane del Petrolio (Italy)
IEOC	International Egyptian Oil Co., Inc. (Egypt)
IFCE	Institut Francais des Combustibles et de l'Energie (France)

IFP	Institut Francaise du Petrole (France)
IGSA	Investigaciones Geologicas, S.A. (Spain)
IIAPCO	Independent Indonesian American Petroleum Co. (Indonesia)
I.L.S.E.A.	Industria Leganti Stradali et Affini (United Kingdom)
I.M.E.	Industrias Matarazzo de Energia (Brazil)
IMEG	International Management & Engineering Group of Britain Ltd. (United Kingdom)
IMEG	Iranian Management & Engineering Group Ltd. (Iran)
IMINOCO	Iranian Marine International Oil Co. (Iran)
IMS	Industria Metalurgica de Salvador, S/Z (Brazil)
I.N.C.I.S.A.	Impresa Nazionale Condotte Industriali Strade Affini (United Kingdom)
INDEIN	Ingenieria y Desarrolio Industrial S.A. (Spain)
INI	Instituto Nacional de Industria (Spain)
INOC	Iraq National Oil Co. (Iraq)
INTERCOL	International Petroleum (Colombia) Ltd. (Colombia)
IODRIC	International Oceanic Development Research Information Center (Japan)

IOE & PC	Iranian Oil Exploration & Producing Co. (Iran)
IORC	Iranian Oil Refining Co. N.V. (United Kingdom)
IPAC	Iran Pan American Oil Co. (Iran)
I.P.L.O.M.	Industria Piemontese Lavorazione Oil Minerali (United Kingdom)
IPLAS	Industrija Plastike (Yugoslavia)
IPRAS	Istanbul Petrol Rafinerisi A.S. (Turkey)
IRANOP	Iranian Oil Participants Limited (England)
IROM	Industria Raffinazione Oil Minerali (Italy)
ITOPCO	Iranian Offshore Petroleum Company (United Kingdom)
IROS	Iranian Oil Services, Ltd. (England)
IVP	Instituto Venezolano de Petroquimica (Venezuela)
JAPEX	Japan Petroleum Trading Co. Ltd. (Japan)
KIZ	Kemijska Industrijska Zajednica (Yugoslavia)
KNPC	Kuwait National Petroleum Co. (Arabia)
KSEPL	Kon./Shell Exploration and Production Laboratory (Netherlands)
KSPC	Kuwait Spanish Petroleum Co. (Kuwait)
KUOCO	Kuwait Oil Co., Ltd. (England)
LPACO	Lavan Petroleum Co. (Iran)
LEMIGAS	Lembaga Minjak Dan Gas Bumi (Indonesia)

LINOCO	Libyan National Oil Corp. (Libya)
L.M.B.H.	Lemgaga Kebajoran & Gas Bumi (Libya)
MABANAFT	Marquard & Bahls B.m.b.H. (Germany)
MATEP	Materials Tecnicos de Petroleo S.A. (Brazil)
MAWAG	Mineraloel Aktien Gesellschaft ag (Germany)
MEDRECO	Mediterranean Refining Co. (Lebanon)
MEKOG	N.B. Maatschappij Tot Exploitatie van Kooksovengassen (Netherlands)
MENEG	Mene Grande Oil Co. (Venezuela)
METG	Mittelrheinische Ergastransport GmbH (Germany)
M.I.T.I.	Ministry of International Trade and Industry (Japan)
MODEC	Mitsui Ocean Development & Engineering Co. Ltd. (Japan)
MPL	Murco Petroleum Limited (England)
NAKI	Nagynyomasu Kiserleti Intezet (Hungary)
NAM	N.V. Nederlandse Aardolie Mij. (Netherlands)
NAPM	N.V. Nederlands Amerikaanse Pijpleiding Maatschappij (Netherlands)
NCM	Nederlandse Constructiebedrijven en Machinefabriken N.V. (Netherlands)

NDSM	Nederlandse Dok en Scheepsbouw Maatschappij (Netherlands)
NED.	North Sea Diving Services, N.V. (Netherlands)
NEPTUNE	Soc. de Forages en Mer Neptune (France)
NETG	Nordrheinische Erdgastransport Gesselschaft mbH (Germany)
NEVIKI	Nehezvegyipari Kutato Intezet (Hungary)
NIOC	National Iranian Oil Co. (Iran)
NORDIVE	North Sea Diving Services Ltd. (United Kingdom)
NOSODECO	North Sumatra Oil Development Cooperation Co. Ltd (Indonesia)
NPC	Nederlandse Pijpleiding Constructie Combinatie (Netherlands)
NPCI	National Petroleum Co. of Iran (Iran)
N.V.A.I.G.B.	N.V. Algemene Internationale Gasleidingen Bouw (Netherlands)
N.V.G.	Nordsee Versorgungsschiffahrt GmbH (Germany)
NWO	Nord-West Oelleitung GmbH (Germany)
OCCR	Office Central de Chauffe Rationnelle (France)
OEA	Operaciones Especiales Argentinas (Argentina)

OKI	Organsko Kenijska Industrija (Yugoslavia)
OMNIREX	Omnium de Recherches et Exploitations Petrolieres (France)
OMV	Oesterreichische Mineraloelverwaltung A.G. (Australia)
OPEC	Organization of Petroleum Exporting Countries
OTP	Omnium Techniques des Transprots par Pipelines (France)
PCRB	Compagnie des Produits Chimiques et Raffineries de Berre (France)
PEMEX	Petroleos Mexicanos (Mexico)
PERMAGO	Perforaciones Marinas del Golfo S.A. (Mexico)
PETRANGOL	Companhia de Petroleos de Angola (Angola)
PETRESA	Petroquimica Espanola, S.A. (Spain)
PETROBRAS	Petroleo Brasileiro S.A. (Brazil)
PETROLIBER	Compania Iberica Refinadora de Petroleos, S.A. (Spain)
PETROMIN	General Petroleum and Mineral Organization (Saudi Arabia)
PETRONOR	Refineria de Petroleos del Norte, S.A. (Spain)
PETROPAR	Societe de Participations Petrolieres (France)
PETROREP	Societe Petroliere de Recherches Dans La Region Parisienne (France)

POLICOLSA	Poliolefinas Colombianas S.A. (Colombia)
PREPA	Societe de Prospection et Exploitations Petrolieres en Alsace (France)
PRODESA	Productos de Estireno, S.A. de C.V. (Mexico)
PROTEXA	Construcciones Protexa, S.A. de C.V. (Mexico)
PYDESA	Petroleos y Derivados, S.A. (Spain)
QUIMAR	Quimica del Mar, S.A. (Mexico)
RAP	Regie Autonome des Petroles (France)
RASIOM	Raffinerie Siciliane Olii Minerali (Esso Standard Italiana S.p.A.) (Italy)
RDM	De Rotterdamsche Droogdok Mij. N.V. (Netherlands)
RDO	Rhein-Donau-Oelleitung GmbH (Germany)
REDCO	Rehabilitation, Engineering and Development Co. (Indonesia)
REPESA	Refineria de Petroleos de Escombreras, S.A. (Spain)
REPGA	Recherche et Exploitation de Petrole et de Gaz (France)
RIOGULF	Rio Gulf de Petroleos, S.A. (Spain)
SACOR	Sociedade Anonima Concessionaria da Rafinacao de Petroleos em Portugal (Portugal)
SAEL	Sociedad Anonima Espanola de Lubricantes (Spain)

S.A.F.C.O.	Saudi Arabian Refinery Co. (Saudi Arabia)
SAFREP	Societe Anonyme Francaise de Recherches et d'Exploitation de Petrole (France)
SAIC	Sociedad Anonima Industrial y Commercial (Argentina)
SAM	Societe d'Approvis de Material Patrolier (France)
SAP	Societe Africaine des Petroles (France)
SAPPRO	Societe Anonyme de Pipeline a Produits Petroliers sur Territoire Genevois (Switzerland)
SAR	Societe Africaine de Raffinage (Dakar)
SARAS	S.p.a. Raffinerie Sarde (Italy)
SARL	Chimie Development International (Germany)
SAROC	Saudi Arabia Refinery Co. (Saudi Arabia)
SAROM	Societa Azionaria Raffinazione Olli Minerali (Italy)
SARPOM	Societa per Azioni Raffineria Padana Olii Minerali (Italy)
SASOL	South African Coal, Oil and Gas Corp. Ltd. (South Africa)
S.A.V.A.	Societa Alluminio Veneto per Azioni (Italy)
SCC	Societe Chimiques des Charbonnages (France)

SCI	Societe Chimie Industrielle (France)
SPC	Societe Cherifienne des Petroles (Morocco)
SECA	Societe Europreeme des Carburants (Belgium)
SEHR	Societe d'Exploitation des Hydrocarbures d'Hassi R'Mel (France)
SEPE	Sociedad de Exploracion de Petroleos Espanoles, S.A. (Spain)
SER	Societe Equatoriale de Raffinage (Gabon)
SERCOP	Societe Egyptienne pour le Raffinage et le Commerce du Petrole (Egypt)
SEREPT	Societe de Recherches et d'Exploitation des Petroles en Tunisia (Tunisia)
SER VIPETROL	Transportes y Servicios Petroleros (Ecuador)
SETRAPEM	Societe Equatoriale de Travaux Petroliers Maritimes (France, Germany)
SFPLJ	Societe Francaise de Pipe Line du Jura (France)
SHELLREX	Societe Shell de Recherches et d'Exploitations (France)
S.I.B.P.	Societe Industrielle Belge des Petroles (Belgium)
SIF	Societe Tunisienne de Sondages, Injections, Forages (Tunisia)
SINCAT	Societa Industriale Cantese S.p.a. (Italy)
SIPSA	Sociedad Investigadora Petrolifera S.A. (Spain)

SIR	Societa Italiana Resine (Italy)
SIREP	Societe Independante de Recherches et d'Exploitation du Petrole (France)
SIRIP	Societe Irano-Italienne des Petroles (Iran)
SITEP	Societe Italo-Tunisienne d'Exploitation (Italy, Tunisia)
SMF	Societe de Fabrication de Material de Forage (France)
SMP	Svenska Murco Petroleum Aktiebolag (Sweden)
SMR	Societe Malagache de Raffinage (Malagasy)
SNGSO	Societe Nationale des Gas de Sud-Ouest (France)
SN MAREP	Societe Nationale de Material pour la Recherche et l'Exploitation du Petrole (France)
SNPA	Societe Nationale des Petroles d'Aquitaine (France)
SN REPAL	Societe Nationale de Recherches et d'Exploitation des Petroles en Algerie (France)
SOCABU	Societe du Caoutchouc Butyl (France)
SOCEA	Societe Eau et Assainissement (France)
SOCIR	Societe Congo-Italienne de Raffinage (Congo Republic)

SOFEI	Societe Francaise d'Enterprises Industrielles (France)
SOGARES	Societe Gabonaise de Realisation de Structures (France)
SOMALGAZ	Societe Mixte Algerienne de Gaz (Algeria)
SOMASER	Societe Maritime de Service (France)
SONAP	Sociedade Nacional de Petroleos S.A.R.I. (Portugal)
SONAREP	Sociedade Nacional de Refinacao de Petroleos S.A.R.L. (Mozambique)
SONATRACH	Societe Nationale de Transport et de Commercialisation des Hydrocarbures (Algeria, France)
SONPETROL	Sondeos Petroliferos S.A. (Spain)
SOPEFAL	Societe Petroliere Francaise en Algeria (Algeria)
SOPEG	Societe Petroliere de Gerance (France)
SOREX	Societe de Recherches et d'Exploitations Petrolieres (France)
SOTEI	Societe Tunisienne de Enterprises Industrielles (Tunisia)
SOTHRA	Societe de Transport du Gaz Naturel D'Hassi-er-r'mel a Arzew (Algeria)
SPAFE	Societe des Petroles d'Afrique Equatoriale (France)
SPANGOC	Spanish Gulf Oil Co. (Spain)

SPEICHIM	Societe Pour l'Equipment des Industries, Chimiques (France)
SPG	Societe des Petroles de la Garrone (France)
S.P.I.	Societa Petrolifera Italiana (Italy)
SPIC	Southern Petrochemical Industries Corporation Ltd.
SPLSE	Societe du Pipe-Line Sud Europeen (France)
SPM	Societe des Petroles de Madagascar (France)
SPV	Societe des Petroles de Valence (France)
SSRP	Societe Saharienne de Recherches Petrolieres (France)
STEG	Societe Tunisienne d'Electricite et de Gaz (Tunisia)
STIR	Societe Tuniso-Italienne de Raffinage (Tunisia)
TAL	Deutsche Transalpine Oelleitung GmbH (Germany)
TAMSA	Tubos de Acero de Mexico, S.A. (Mexico)
TATSA	Tanques de Acero Trinity, S.A. (Mexico)
TECHINT	Compania Technica Internacional (Brazil)
TECHNIP	Compagnie Francaise d'Etudes et de Construction Technip (France)
TEXSPAIN	Texaco (Spain) Inc. (Spain)
TORC	Thai Oil Refinery Co. (Thailand)
T.P.A.O.	Turkiye Petrolleri A.O. (United Kingdom)

TRAPIL	Societe des Transports Petroliers Par Pipeline (France)
TRAPSA	Compagnie des Transports par Pipe-Line au Sahara (Algeria)
UCSIP	Union des Chambres Syndicales de l'Industrie du Petrole (France)
UPG	Union Generale des Petroles (France)
UIE	Union Industrielle et d'Enterprise (France)
UNIAO	Refinaria e Exploracao de Petroleo "UNIAO" S.A. (Brazil)
URAG	Unteweser Reederei Gmbh (Germany)
URG	Societe pour l'Utilisation Rationnelle des Gaz (France)
WEPCO	Western Desert Operating Petroleum Company (Egypt)
YPF	Yacimientos Petroliferos Fiscales (Argentina)
YPFB	Yacimientos Petroliferos Fiscales Bolivianos (Bolivia)

API STANDARD OIL-MAPPING SYMBOLS

Location .... ⭕

Abandoned location ....erase symbol

Dry hole ....

Oil well .... ●

Abandoned oil well ....

Gas well ....

Abandoned gas well ....

Distillate well ....

Abandoned distillate well ....

Dual completion—oil .... ◉

Dual completion—gas ....

Drilled water-input well .... ⌀ W

Converted water-input well .... ● W

Drilled gas-input well .... ⌀ G

Converted gas-input well .... ● G

Bottom-hole location .... ⭕---x
    (x indicates bottom of hole. Changes in well
    status should be indicated as in symbols
    above.)

Salt-water disposal well .... ⊕ SWD

Courtesy American Petroleum Institute, Division of Production.

MATHEMATICAL SYMBOLS AND SIGNS

+	plus	∴	therefore
−	minus	∵	because
±	plus or minus	:	is to; divided by
×	multiplied by	::	as; equals
·	multiplied by	∷	geometrical proportion
÷	divided by	∝	varies as
/	divided by	≐	approaches a limit
=	equal to	∞	infinity
≠	not equal to	∫	integral
≈	nearly equal to	d	differential
≅	congruent to	∂	partial differential
≡	identical with	Σ	summation of
≢	not identical with	!	factorial product
⇔	equivalent to	π	pi (3.1416)
>	greater than	e	epsilon (2.7183)
≯	not greater than	°	degree
<	less than	′	minute; prime
≮	not less than	″	second
≧	greater than or equal to	∠	angle
≦	less than or equal to	∟	right angle
∼	difference between	⊥	perpendicular
≎	difference between	○	circle
−:	difference between	⌒	arc
√	square root	△	triangle
∛	cube root	□	square
ⁿ√	nth root	▭	rectangle

GREEK ALPHABET

A	α	Alpha	N	ν	Nu
B	β	Beta	Ξ	ξ	Xi
Γ	γ	Gamma	O	o	Omicron
Δ	δ	Delta	Π	π	Pi
E	ε	Epsilon	P	ρ	Rho
Z	ζ	Zeta	Σ	σ	Sigma
H	η	Eta	T	τ	Tau
Θ	θ	Theta	Υ	υ	Upsilon
I	ι	Iota	Φ	φ	Phi
K	κ	Kappa	X	χ	Chi
Λ	λ	Lambda	Ψ	ψ	Psi
M	μ	Mu	Ω	ω	Omega

METRIC-ENGLISH SYSTEMS
CONVERSION FACTORS

Basic Dimensions

Metric System

Length—   meter (m)
              kilometer (km)
              centimeter (cm)
              millimeter (mm)

Area—   square meters (m^2)
              square centimeters (cm^2)

Volume—   cubic meters (m^3)
              cubic centimeters (cm^3)
              liters (l)
              milliliters (ml)

Mass—   kilograms (kg)
              grams (g)
              gram-moles (gm-moles)
              kilogram-moles (kg-moles)

Density— kg/m^3, g/cm^3

English System

Length—   inch (in.)
              foot (ft)
              yard (yd)
              mile (mile)

Area—   square inches ($in.^2$)
              square feet (ft^2)
              square miles ($miles^2$)

Volume—   cubic inches ($in.^3$)
              cubic feet (ft^3)
              barrels (bbl)
              U.S. gallons (gal)
              Imperial gallons (Imp. gal)

Mass—   pounds (lb)
              pound-moles (lb-moles)

Density—   pounds per gallon (lb/gal,
                lb/ft^3)

System Equivalents

Metric System

Length—	1 m = 100 cm = 1,000 mm = 0.001 km
Area—	1 m² = 10,000 cm²
Volume—	1 m³ = 1,000,000 cm³ = 1,000,000 ml = 1,000 l
Mass—	1 kg = 1,000 g
	1 kg-mole = 1,000 gm-moles
Density—	1 kg/m³ = 0.001 g/cm³

English System

Length—	1 ft = 12 in. = 0.333 yd = 0.000189 miles
	1 mile = 5,280 ft = 1,750 yd
Area—	1 ft² = 144 in.²
	1 mile² = 27,878,400 ft²
Volume—	1 ft³ = 1,728 in.³ = 0.178 bbl = 7.48 U.S. gal = 6.23 Imp. gal
	1 bbl = 5.61 ft³ = 42 U.S. gal = 34.97 Imp. gal
Density—	1 lb/gal = 7.48 lb/ft³ = 42 lb/bbl

BASIC CONVERSION FACTORS

Length— 1 m = 3.281 ft = 39.37 in.
1 ft = 0.305 m = 30.5 cm =
3,050 mm
1 mile = 1.61 km
1 km = 0.621 mile

Area— $1 m^2$ = 10.76 ft^2 = 1,549 $in.^2$
$1 ft^2$ = 0.0929 m^2 = 929.4
cm^2

Volume— $1 m^3$ = 35.32 ft^2 = 6.29 bbl
1 l = 0.035 ft^3 = 61 $in.^3$
$1 ft^3$ = 0.0283 m^3 = 28.31
1 bbl = 0.159 m^3 = 1591

Mass— 1 kg = 2.205 lb
1 lb = 0.454 kg = 454 g
1 metric ton = 1,000 kg =
2,205 lb

Density— $1 kg/m^3$ = 0.0624 lb/ft^3
$1 lb/ft^3$ = 16.02 kg/m^3 =
0.01602 g/cm^3
$1 g/cm^3$ = 62.4 lb/ft^3

Force— 1 kg force = 2.205 lb force
1 lb force = 0.454 kg force

Work & Heat— 1 Btu = 0.252 kilocalories
(kcal)
1 kcal = 3.97 Btu

Power— 1 kilowatt (kw) = 860 kcal/
hr = 3,415 Btu/hr =
1.341 horsepower (hp)
1 hp = 0.746 kw = 641 kcal/
hr = 2,545 Btu/hr

Enthalpy— 1 kcal/kg = 1.8 Btu/lb
1 Btu/lb = 0.556 kcal/kg

Pressure— 1 bar = 14.51 $lb/in.^2$ (psi) =
0.987 atmospheres (atm)
= 1.02 kg/cm^2

BASIC CONVERSION FACTORS *(cont'd.)*

$$1 \text{ kg/cm}^2 = 14.22 \text{ psi} = 0.968 \text{ atm}$$

$$1 \text{ psi} = 0.0703 \text{ kg/cm}^2$$

Temperature—
$$°C = 0.556 \, (°F - 32)$$
$$°K = °C + 273$$
$$°F = 1.8°C + 32$$
$$°R = °F + 460$$

Common Oilfield Spellings

A
about-face
aboveground
acknowledgment
acre-feet
aeration
aftercooler
aftertreat
afterwash
air flow (n, adj.)
airfoil
air line
airtight
alumna - s fem.
alumnae - pl fem.
alumnus - s masc.
alumni - pl masc.
anti-icer

B
backflow
backlight
back off
back pressure
backup (adj.)
backwash
backwater
base line
baseplate
behavior
belowground
bench mark
bench scale
blow-by
blowdown
blowout
boil off
borehole
bottom hole

breakaway
breakdown
break-even (adj)
break even (v)
breakout
breakpoint
breakthrough
breakup (n)
break up (v)
briquette
bubble cap
buildup (n)
build-up (adj)
build up (v)
built in
burn-off
burnout (n)
burnup (n)
bypass
byproduct

C
Caribbean
carry-over
casinghead
center line
changeable
change out (v)
changeover
channeling
charge stock
checklist
check-out (adj)
check out (v)
checkpoint
city-wide
classroom
cleanout
cleanup (n)

clean up (v)
cleanup (adj)
clear-cut
close up (v)
close-up (n, adj)
closeout
coastline
commitment
commingle
controlling
controlled
co-op
co-owner
coproduct
counterbalance
counterbattery
countercurrent
counterflow
country-wide
crisscross
criterion - **s**
criteria - pl
cross-bedding
cross-country
cross-flow
crosshead
crossover
cross plot
cross-reference
cross section
cutdown
cutoff
cutout
cut point
cycle oil

D

datum - **s**
data - pl
deadman

dead time
deadweight
deepwater (adj)
deep water (n)
de-ethanizer
desiccant
desirable
dew point
doghouse
dogleg
double-jointed
downdip
downdrag
downflow
downgrade
downhole
downstream
downthrown
downtime
drawdown
drawoff (n)
drawworks
drill bit
drill collar
drillhead
drillpipe
drillship
drillsite
drillstem
drillstock
drillstring
drumhead

E

edgewise
end point
en route

F

face-lifting

fail-safe
falloff
farmin (n)
farmout (n)
feedback
feed rate
feedstock
feedwater
fiberglass
fieldman
fill-up
fireflood
firebox
fire wall
firewater (n)
flatbed
flier
flow-control
flow line
flowmeter
flywheel
foam glass
follow-up
forklift
formula
formulas
fourfold
freeze-up
freshwater (adj)
frost line
full time

G

gamma ray
gas oil
gearbox
gearshift
grassroots
gray
ground line

groundwater
guesswork
guideline

H

halfway
hammerblow
handwritten
hardback
headlong
hold-down
holdup
homemade
hookup

I

in between (v)
in-between (n, adj)
industry-wide
infill
inflow (n)
in-line
inrush
iso-octane
   (no other "iso"
   words)

J

jackknife
jackup (n)
jackup (n)
jack up (v)
jobsite
judgment

K

kerosene
know-how
knockout
knowledgeable

L

landfill
landmass
lay barge
laydown (n)
lay-down (adj)
lay down (v)
leakoff
leakproof
left-hand
lengthways
letdown
lightweight
lineup (n)
line up (v)
linkup (n)
link up (v)
lowboy
lockout

M

main line
mainstream
makeup
manageable
man-day
man-hour
man-year
mathematics
measurable
Mediterranean
Midcontinent
Mideast
midyear
mile-wide
millsite
multimillion-dollar
minable
modeling
mountainside

mousehole
movable
mud cake
mud line

N

nationwide
nearby
non-Communist
noticeable

O

observable
oceanfront
offgas
off-line
off-loading
off site
offtake
oil field (n)
oilfield (adj)
oilman
oil well (n)
oil-well (adj)
onshore
on site (n)
on stream (n)
open hole
outfall
overall
overhead
overpressure
overnight
override

P

paperback
passthrough
payout
percent

phaseout
pickup
piggyback
pinchout
pipelay
pipelaying
pipelayer
pipeline
pipe rack
pipe still
powerhouse
predominant
printout
proof-test
pullout (n, adj)
pull out (v)
push button

R

rainwater
rathole
readout
realizable
real time
reconnaissance
reentry
removable
right-of-way
rig-up (n)
rig up (v)
ringwall
riprap
roundoff
round trip
runback
rundown (n)
run down (v)
run-down (adj)
run forward
runoff

S

salable
saltwater (adj)
salt water (n)
salvageable
sandblasting
scale-up
seabed
sealift
seawater
second hand (n)
secondhand (adj)
sendout
set point
setup (n)
severalfold
shipshape
shipside
shipyard
shoreline
shortcut
shortsighted
shortwave
shot hole
shot point
shut down (v)
shutdown (n, adj)
shut in (v)
shutin (n, adj)
shutoff (adj)
side boom
side cut
side draw
side stream
sidetrack
sidewall
sizable
slackline
slip joint
slipstream

slow-up
soleplate
spanwise
spectrum - **s**
spectra - pl
spot-check
standby
standoff
standpipe
standpoint
standpost
standstill
start-up (n, adj)
start up (v)
steamflood
steam line
step-out (n)
step-up (n, adj)
stepwise
straightforward
straight-run
stratum - **s**
strata - pl
stiffleg
stopgap
subpoena
superheat
switchgear

T

tailor-made
takeoff
takeover
take-up
tank car
tank truck
teardown
tie-in
tie-down
titleholder

toolpusher
toss-up
towline
trademark
trade name
trade off
trade out
transatlantic
traveling
trouble-free
troubleshoot
trunk line
turn down (v)
turndown (n, adj)
turnkey

U

ultraviolet
underwater
underway
updip
upflow
upstream
up-to-date
usable

W

warm-up (adj)
washout
waste water
water-cooled
water cut
waterflood
waterhead
waterline
waterside
watertight
waveform
wave front
wavelength

weathertight
wellhead
wellbore
wellsite
wellstream
wet out
whipstock
windblow
wind-chill (adj)
wireline
workboat
workbox

work load
workover (n, adj)
work over (v)
worldwide
wraparound

X

X-ray

Y

year-end

ADDENDUM

ADDENDUM

ADDENDUM

ADDENDUM

ADDENDUM

ADDENDUM

ADDENDUM

ADDENDUM

ADDENDUM

ADDENDUM

ADDENDUM

ADDENDUM